Autorenkollektiv

Organotin Compounds

Structure and Reactivity of Monomeric, Molecular Tin(II) Compounds

M. Veith, O. Recktenwald

Chirality, Static and Dynamic Stereochemistry of Organotin Compounds

M. Gielen

Coordination Effects in Formation and Cross-Linking Reactions of Organotin Macromolecules

Z. M. O. Rzaev

AKADEMIE-VERLAG · BERLIN

Autorenkollektiv

Organotin Compounds

Autorenkollektiv

ORGANOTIN COMPOUNDS

With 34 Figures and 22 Tables

AKADEMIE-VERLAG · BERLIN
1983

Die Originalausgabe erscheint im Springer-Verlag Berlin · Heidelberg · New York

Vertrieb ausschließlich für die DDR und die sozialistischen Länder

Erschienen im Akademie-Verlag, DDR-1086 Berlin, Leipziger Straße 3–4
Lizenznummer: 202 · 100/550/82
Gesamtherstellung: VEB Druckerei „Thomas Müntzer“, 5820 Bad Langensalza
Umschlaggestaltung: Karl Salzbrunn
Bestellnummer: 763 186 2 (6741) · LSV 1275
Printed in GDR
DDR 76,— M

Table of Contents

Structure and Reactivity of Monomeric, Molecular Tin(II) Compounds

Michael Veith and Otmar Recktenwald

Institut für Anorganische Chemie der Technischen Universität, Braunschweig, FRG

Table of Contents

1 Introduction

As it is well-known, the element tin is often used as a hetero-atom in organic chemistry as well as in organo-metallic or molecular inorganic chemistry. Many reactions can be performed with the assistance of this element as described in W. P. Neumann's classical book [1]. While inorganic chemists have been familiar from the very beginning with the two possible stable oxidation states of tin (+II and +IV), the main activity in molecular chemistry has for a long time been focused on the oxidation state +IV. Only in the last decade has considerable attention been payed to molecular tin(II) compounds and extensive progress has been performed in this field. This fact is also well illustrated in review articles: fifteen years ago only a small number of them were known, the most relevant being those of O. M. Nefedov and M. N. Manakov [2] and especially of J. D. Donaldson [3]. In the meantime, new results of studies on the structure of tin compounds have been collected and discussed by P. G. Harrison [4], J. A. Zubieta and J. J. Zuckerman [5]; a bibliography on the structures has been published by P. A. Cusack et al. [6], and the syntheses and reactions of some organic derivatives of tin(II) have been compared by W. P. Neumann [7] and J. W. Connolly and C. Hoff [8]. The electronic properties of carbene analogs have been analyzed by O. M. Nefedov et al. [9] while ring and cage compounds have been described by M. Veith [10].

This article is an attempt at evaluating new important features of tin(II) chemistry: the central point is the interrelationship between molecular structure and reactivity of molecular tin(II) compounds. To define these compounds more closely, only those are discussed which are stable, monomeric in solvents and which may be classified as carbene analogs [2]. Thus, not a complete survey of tin(II) chemistry is given but stress is laid on the structures and reactions of selected compounds. A general introduction to the subject precedes the main chapters. For comparison, also solid-state tin(II) chemistry is included to demonstrate the great resemblance with molecular tin(II) chemistry. Tin(II) compounds, which are either generated as intermediates or only under definite conditions such as temperature or pressure, are not described in detail.

2 General Aspects of Low-Valency Elements

2.1 The "Inert s-Pair Effect"

In books on inorganic chemistry, the marked increase in the stability of the lower oxidation state (by two units) of heavier elements descending the main groups of the periodic Table is often explained by the "inert s-pair effect" (see J. E. Huheey [11]). For example, elements like In and Sn may use only 1 or 2 electrons for the formation of bonds instead of 3 or 4 (group number), leaving one electron pair in the outer valence shell "inert". The electron pair is assumed to occupy an s-orbital. This classification does not very much contribute to the understanding of bonding; first

Table 1. Ionization energies (eV) of s electrons [11]

Element	$IE_2 + IE_3$	Element	$IE_3 + IE_4$
B	63.1	C	112.1
Al	47.2	Si	78.6
Ga	51.2	Ge	79,9
In	46.9	Sn	71.2
Tl	50.2	Pb	74.2

of all, it is not certain whether the non-bonding electron pair really occupies a pure s-orbital (see also in Chapter 3). Secondly, when comparing the ionization energies of group III and IV elements (Table 1), there is no marked increase in the stability of s-electrons with rising atomic weight; a slight effect may be noticed with the elements Tl and Pb.

J. E. Huheey [11] has proposed another explanation following the reasoning of R. S. Drago [12]. If the enthalpies of the reaction

$$EX_4 \rightarrow EX_2 + X_2$$

are compared using some halides of group IV elements (Table 2), it is quite obvious that the reaction proceeds more easily with heavier elements (enthalpies becoming less positive!). This effect is reflected by a gain of bond energy when the element passes from the higher oxidation state to the lower one (Table 3). This, of course, means that the halogen atoms are more strongly bound to the metal atom in the low-valence state. Disregarding the different electronegativities in the two oxidation states, two reasons may account for an extra stabilization of the bonds at lower oxidation number of the group IV element:

1) in the higher valence state the promotion energy $s^2p^2 \rightarrow sp^3$ must be compensated;

2) heavier elements suffer from inner electron repulsions, which may weaken the bonds.

When changes in electronegativity are taken into account (the element passing from the higher oxidation number to the lower one) another stabilization effect should be added to the two already mentioned. As can be seen in Chapter 3, the element in the lower oxidation state is always more electropositive: the covalent bond between the central atom and the usually more electronegative ligand will become more polar and will therefore be reinforced by a superimposed ionic component. On the other side, the bond lengths in comparable compounds are longer for lower oxidation states!

Table 2. Enthalpies (kJ/mol) of the reaction: $EX_4 \rightarrow EX_2 + X_2$ [11]

X=	F	Cl	Br	J
Element E				
Ge	+695	+381	+260	+167
Sn	+544	+276	+243	+142
Pb	+385	+121	+ 88	+ 17

Table 3. Bond energies (kJ/mol) of EX_2 and EX_4 [11]

Compound	EF_2	EF_4	ECl_2	ECl_4
Element E				
Ge	481	452	385	354
Sn	481	414	386	323
Pb	394	331	325	243

To sum up, it is clear that havier elements tend to achieve an oxidation number which is by two units lower than the group number. On the other side, a straightforward explanation for this phenomenon is hard to find.

2.2 Carbenes and Carbene Analogs

A very useful class of intermediates in synthetic organic chemistry are carbenes $|CX_2$ (X = H, R, F, Cl, Br etc.) [13]. These molecules are typical representatives of molecular, monomeric, highly reactive and electronically unsaturated compounds. Their main characteristics are the following:

1) there are only six electrons in the valence shell;
2) one electron pair is non-bonding.

The non-bonding electron pair may occupy one orbital with antiparallel spins (singlet, $^1\sigma^2$), or two different orbitals with antiparallel (singlet, $^1\sigma p$) or parallel spins (triplet, $^3\sigma p$).

It has become common to classify all *molecular* compounds, which fulfill the above characteristics, as carbene analogs [9,13]. As a consequence, compounds of divalent silicon, germanium, tin, and lead may be regarded as carbene-like and are therefore called silylenes, germylenes, stannylenes, and plumbylenes. In contrast to carbenes they have one property in common: the energetically most favorable electronic state is the singlet $^1\sigma^2$ found by experiments and calculations [9].

There are two possibilities to describe the structure of these singlet carbene analogs:

1) The non-bonding electron pair occupies an s-orbital, the bonding electrons occupy p-orbitals, while the third p-orbital remains empty. The bonding angle should be strictly 90° (geometry A).
2) Both non-bonding and bonding electron pairs occupy sp^2-hybrid orbitals while again a p-orbital is unoccupied. In this case, the bonding angle should be 120° (geometry B).

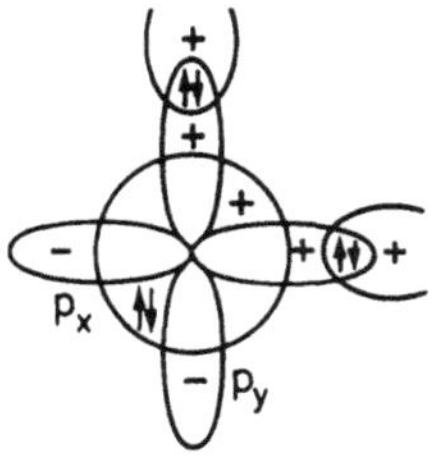

A

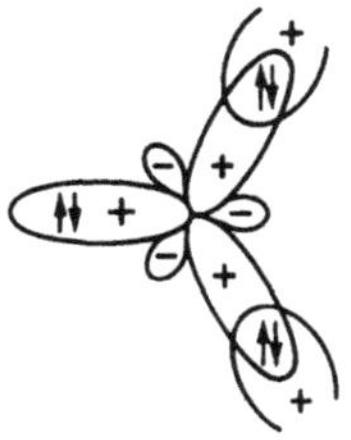

B

Table 4. Bond angles (°) in EX_2 [9)]

Compound	Angle X—E—X	Compound	Angle X—E—X
SiF_2	100–101	SnF_2	92–95
$SiCl_2$	104–105	$SnCl_2$	95
GeF_2	94–97	PbF_2	90–95
$GeCl_2$	95–107	$PbCl_2$	95

When experimentally determined bond angles of carbene analogs are compared, the values are found to be in between these two possibilities; Table 4 lists the bond angles of some halides in the vapor phase as determined mostly from electron diffraction or vibrational and rotational spectra. With increasing atomic number of the central atom, the angles seem to approach 90°. This may be explained by either a steric effect or the more pronounced hybridization in lighter elements. Referring to the theoretical angles of models A and B the deviation found is often accounted for by repulsion forces between the ligands (model A) or repulsion between the lone pair and the ligands in model B [14)]. For heavy elements model A seems to be more important, and we should expect divalent tin compounds to have this geometry. As can be seen in Chapter 4, this is the case for all tin compounds known so far. At the same time, it should be concluded that the non-bonding electron pair at the central tin atom exhibits no stereochemical activity because it is located in a radially distributed s-orbital. Structures of solid tin(II) compounds clearly demonstrate the contrary (see Chapter 3). We must therefore assume that the s-orbital must be mixed with energetically favorable orbitals, allowing a deviation from a spherical shape [15,16)].

It should be clear by the definition given so far that the carbene-analogous state is limited to molecular species. The oligomer of EX_2 $(EX_2)_n$ is, of course, much more stable than $|EX_2$ in every respect. It should nevertheless be noted that also the oxidation number does not change in going from the monomer to the polymer the chemical, structural, and electronic properties of these species are completely different.

2.3 Stabilization of Carbene Analogs

The stability of molecules depends in the first place on limiting conditions. Small, mostly triatomic silylenes and germylenes have been synthesized successfully at high temperatures and low pressures [17,18)]. Their reactions can be studied by warming up the frozen cocondensates with an appropriate reactant, whereas their structures are determined by matrix techniques [17,18)]. In addition, reactions in the gas phase or electron diffraction are valuable tools for elucidating the structures and properties of these compounds. In synthetic chemistry, adequate precursors are often used to produce intermediates which spontaneously react with trapping reagents [7)]. The analysis of the products is then utilized to define more accurately the structure of the intermediate.

Neither of these two methods can be employed for the synthesis of stable carbene analogs at ambient temperature. Thus, the only possibility of preparing stable molecules is to modify the properties of the carbene analogs by substitution at the central atom. Two general aspects seem to be important for all electron-deficient compounds:

1) lack of electrons at the central atom has to be compensated by electron-releasing groups;

2) the substituent should be as bulky as possible to prevent polymerization.

The second point is certainly the most important as seen in the case of the stannylenes (Chapter 4) while the first point needs some further discussion. The electron-releasing substituent may act via a simple inductive σ-effect (A) or via a mesomeric π-effect (B), or via both effects.

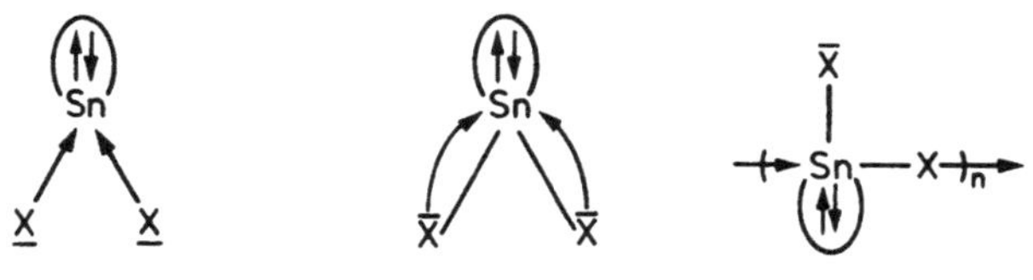

The intramolecular Lewis acid-base interaction of type B is of course always in competition with an intermolecular interaction, as indicated by formula C. Again, a bulky group in α-position to X can favor the formation of monomer B.

3 General Structural Aspects of Tin(II) Chemistry

3.1 Comparison between Structures of Sn(II) and Sn(IV) Compounds

The major difference between structures of tin(II) and tin(IV) compounds (which may be ionic or covalent) can be seen in the coordination sphere of the tin atom. While tin(II) compounds are mostly bent, pyramidal or otherwise distorted, tin(IV) compounds adopt regular geometries as tetrahedra, bipyramides or octahedra, depending on the coordination number. The reason for this phenomenon may readily be explained by the different electronic states: while in tin(IV) compounds all outer electrons of the tin atom are engaged in bonding, tin(II) compounds have one electron pair that does not participate in bonding (see Chapter 2) and displays stereochemical activity. We can consider this electron pair as a further ligand in the coordination sphere of the tin atom. There are only few tin compounds where the lone pair is not stereochemically active. For example, in the cubic form of SnSe and in SnTe the tin atom is situated in the middle of a regular octahedron [19]. It is remarkable that in these cases the cation is surrounded by easily polarizable anions.

According to Table 5, the electronegativity of tin changes with its oxidation number, the lower oxidation state being connected with a more electropositive character. It may be concluded that tin(II) compounds are more ionic than the corresponding tin(IV) derivatives or — from another point of view — more electrophilic and thus more tightly coordinated by surrounding nucleophilic ligands.

While the covalent radius of Sn(IV) can easily be derived from grey tin (r(Sn(IV)) = 140 pm), it is much more difficult to evaluate the radius r(Sn(II)). R. E. Rundle and D. H. Olson [20] proposed a value which is by 15–20 pm larger than that of tetravalent tin on the basis of solid tin(II) compounds. Since it is difficult to estimate the crystal effect in solid-state structures, we have compiled bond lengths of the corresponding tin compounds in Table 6 which have been determined, with the exception of one case, by electron diffraction measurements in the gas phase. Thus, the covalent radius of Sn(II) r(Sn(II)), can roughly be evaluated as 150 ± 3 pm from this comparison. The larger r(Sn(II)) radius as compared with r(Sn(IV)) can be explained either by a repulsion effect of the lone electron pair at the tin atom or by a weaker σ-bond compared with tin(IV), the tin(II) atom utilizing exclusively p-orbitals for bonding.

Table 5. Comparison of electronegativities (EN) [11]

Oxidation State	EN (Pauling)	EN (Sanderson)
Sn (II)	1.80	1.58
Sn (IV)	1.96	2.02

Table 6. Bond distances Sn—X (pm) of molecules SnX_2 and SnX_4 in the gas phase

Molecule SnX_2	Sn—X	Molecule SnX_4	Sn—X	Ref.
SnR_2 *)	2.28	SnR_4 *)	2.17	21)
$SnCl_2$	2.42–2.43	$SnCl_4$	2.28–2.31	22,23,24,25)
$SnBr_2$	2.55	$SnBr_4$	2.44	22,23,26)
SnI_2	2.73–2.78	SnI_4	2.64	22,23,26)

* R = organic ligand in the crystal

3.2 The Stereochemical Effect of the Lone Electron Pair at Tin(II)

In the preceding chapter it was shown that the coordination sphere around the tin atom is of paramount importance for the structure of tin(II) compounds. J. D. Donaldson [3] was the first to convincingly demonstrate that the coordination polyhedra adopted by the tin atom are almost the same for a definite coordination number of the central atom despite the nature of the ligands attached to the latter. This means that a tin compound which might be classified as ionic displays a structure similar to a compound containing almost covalent bonds as far as the coordination geometry around the tin atom is concerned. We can illustrate this phenomenon with some exemplary compounds of opposite bonding type. At the same time we will generalize this approach and demonstrate the dependence of the bond length Sn—L on the coordination number n for almost any molecular or solid-state SnL_n compound.

3.2.1 Coordination Polyhedra

In Fig. 1 are assembled the most important coordination polyhedra of tin(II) compounds omitting high coordination numbers which are difficult to illustrate. In Table 7 the corresponding bond angles characterizing the deviation from ideal

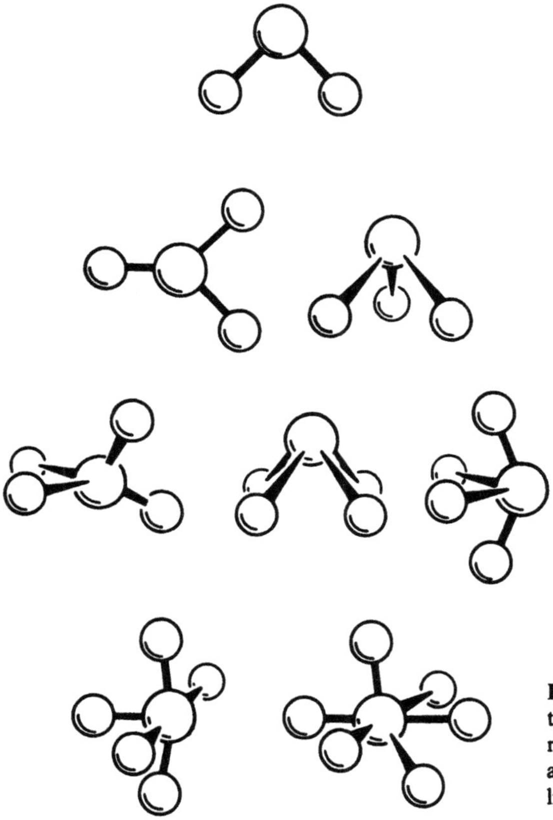

Fig. 1. Coordination polyhedra of the tin(II) atom for coordination numbers n = 2–6. The large circles represent tin atoms, while the smaller ones denote ligands (see also Table 7)

Table 7. Coordination numbers of important tin(II)-ligand arrangements

c.n.	Coordination number	Angles found (°)	Most probable angles (°)
2	ψ-trigonal planar	73–96	90–95
3	trigonal planar	$\Sigma = 360$	95; 2 × 132.5
	trigonal pyramidal	75–95	80–85
4	tetrahedral (distorted)	95–114	109
	tetragonal pyramidal	70–90	75
	ψ-trigonal bipyramidal	130–165;	157
		80–95;	90
		4 × (60–86)	4 × 82
5	ψ-octahedral		
6	distorted octahedral		

geometry or defining the pyramide are listed. The values have been taken from structures compiled in Tables 8–11.

The lowest possible coordination number 2 can only be realized in the ease of tin(II) compounds which have bulky substituents preventing aggregation in the

Table 8. Mean Sn—N distances for different coordination numbers

c.n.	Sn—N (pm)	c. Sphere and mean angle (°)	Compound	Ref.
2	209	bent; 73.2 (cycle!)	$Sn(NCMe_3)_2SiMe_2$	27)
2	209	bent; 96	$Sn[N(SiMe_3)_2]_2$	38)
3	220	trigon. pyram., 81.3	$[Sn(NSiMe_3)_2BMe]_2$	39)
3	222	trig. pyram., 80.5	$(Me_3CN)_3(Me_3AlO)\ Sn_4$	32)
3	223	trig. pyram., 81.3	$Sn_3(NCMe_3)_4H_2$	40)
3	224	trig. pyram., 86.5	$[Sn(NCMe_3)_2SiMe_2]_2$	27)
4	225	squa. pyram., 76	phthalocyaninetin(II)	35)

Table 9. Mean Sn—O distances for different coordination numbers

c.n.	Sn—O (pm)	c. Sphere and mean angle (°)	Compound	Ref.
2	200	bent; 80.8	$Sn[O{-}C_6H_2(CMe_3)_2Me]_2$	28)
3	213	trig. pyram., 89.6	$Sn_3O(OH)_2SO_4$	41)
3	214	,, 83.5	$Ca[Sn(O_2CMe)_3]_2$	42)
3	214	,, 86.3	$Sn_2(OH)PO_4$	43)
3	215	,, 86.1	$Sn_3(PO_4)_2$	44)
3	215	,, 90	$SnFPO_3$	45)
3	216	,, 92	$Sn_{10}W_{16}O_{46}$	46)
3	217	,, 84.4	$SnHPO_3$	47)
3	217	,, 83.3	$KSn(O_2CCH_2Cl)_3$	48)
4	221	squa. pyram., 75	SnO	3)
4	221	ψ-trig. bipyram.	$Sn[OC(Ph)CHC(Me)\ O]_2$	37)
4	223	,,	$Sn[OCH(Ph)CH_2CH(Ph)O]_2$	49)
4	223	squa. pyram., 76	$Sn_6O_4(OMe)_4$	50)
4	224	ψ-trig. bipyram.	$Sn_{10}W_{16}O_{46}$	46)
4	224	squa. pyram., 76	$K_2Sn(C_2O_4)\cdot H_2O$	51)
4	227	ψ-trig. bipyram.	$Sn(O_2CH)_2$	52)
4	227	,,	$Sn_3O(OH)_2SO_4$	41)
4	228	,,	α-$SnWO_4$	36)
6	244	,,	$[Sn^{II}Sn^{IV}(O_2CC_6H_4NO_2)_4O, THF]_2$	53)
6	251	dist. octahedr.	β-$SnWO_4$	54)
7	266	—	$Sn_2(EDTA)\cdot 2\ H_2O$	55)
8	269	—	$Sn(H_2PO_4)_2$	56)
8	272	—	$SnHPO_4$	47,57)
12	293	—	$SnSO_4$	58)

Table 10. Mean Sn—F distance for different coordination numbers

c.n.	Sn—F (pm)	c. Sphere and mean angle (°)	Compound	Ref.
2	191	bent, 94	SnF_2 (gas)	59)
3	208	trig. pyram. 82.4	NH_4SnF_3	31)
3	211	,, 83.3	SnF_2 (monoclinic)	60)
3	215	,, 80.3	Sn_2F_3Cl	61)
3	215	,, 81	Sn_3F_5Br	62)
4	215	ψ-trig. bipyram.	$KSnF_3 \cdot \frac{1}{2} H_2O$	63)
4	217	,,	$Na_4Sn_3F_{10}$	64)
4	218	,,	$Sn_5F_6Br_4$	65)
4	223	,,	$NaSn_2F_5$	66)
5	228	—	SnF_2 (orthorhombic)	67)

Table 11. Mean Sn—S distances for different coordination numbers

c.n.	Sn—S (pm)	c. Sphere and mean angle (°)	Compound	Ref.
3	257	trig. pyram., 95	$BaSnS_2$	68)
3	266	,, , 90.7	SnS	20)
3	267	,, , 86	$Sn^{II}Sn^{IV}S_3$	69)
4	269	ψ-trig. bipyram.	$Sn(S_2C{-}NEt_2)_2$	70, 71)
4	271	,,	$Sn(S_2C{-}OMe)_2$	72)
5	291	—	$Sn[S_2P(OC_6H_5)_2]_2$	73, 74)
7	302	—	$Sn_2P_2S_6$	75)
8	309	—	$Sn_2P_2S_6$	75)
8	311	—	$Sn_4^{II}Sn^{IV}Sb_2S_9$	76)

crystal. Actually, only two compounds are known to display this geometry in the solid state [27, 28]:

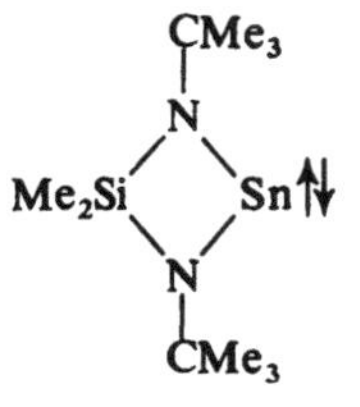

1

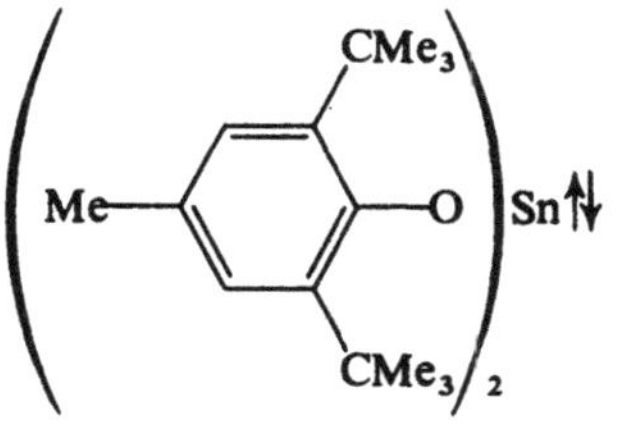

2

The structure of these typical stannylenes *1* and *2* are discussed in Chapter 4 as well as that of bis(pentamethylcyclopentadienyl)tin(II) (*3*) which does not aggregate either but contains two η^5-bonded cyclopentadienyl ligands [29], the coordination number being difficult to establish.

For the coordination number 3, two different environments of tin(II) can be distinguished. One is the trigonal planar arrangement which is realized when the non-bonding electron pair at the tin atom is engaged in bonding, with tin acting as a Lewis base. The first example of this kind characterized by X-ray structural analysis is compound *4* [30] (for the structure see also Chapter 5).

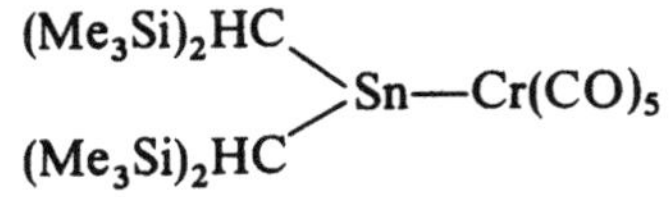

4

No ionic species with this special ligand arrangement around the tin atom has been reported so far.

Most tin(II) compounds display structures with a trigonal pyramidal coordination. This is of course to be expected as the tin atom is in the first place electrophilic in order to complete its outer electron configuration (cf. Chapter 5 and 6). To illustrate the resemblance of this geometry between ionic and molecular compounds, the structure of NH_4SnF_3 (*5*) [31] is compared with that of the cage compound $(Me_3CN)_3(Me_3AlO)Sn_4$ (*6*) [32]. The coordination sphere of the tin atom is the same in *5* and *6* (for the complete structure of *6* see Sect. 6.5):

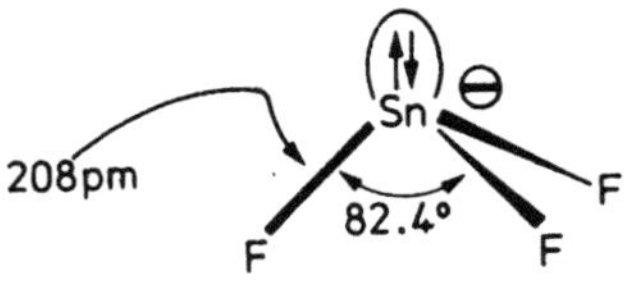

c.f. of 5

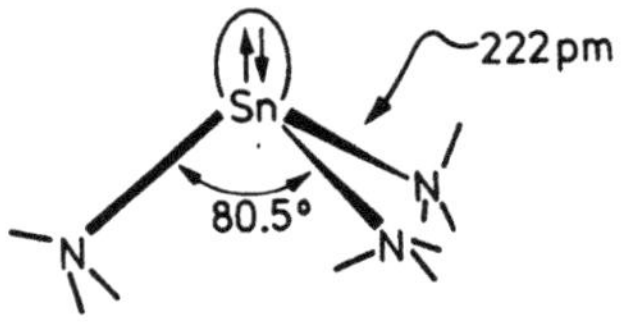

c.f. of 6

(c.f. stands for coordination fig.)

For the coordination number 4 three types of polyhedra have to be considered: a tetrahedron (distorted), a tetragonal pyramide, and a closely related Ψ-trigonal bipyramide.

The first coordination sphere is a special case. It can be generated from the trigonal planar arrangement by adding a further ligand, resulting a tin atom which simultaneously acts as an acid and a base. An illustrative example for this kind of bonding is compound *7* [33] in which the tin atom receives electrons from pyridine and transfers electrons to the chromium atom (see also Chapter 6).

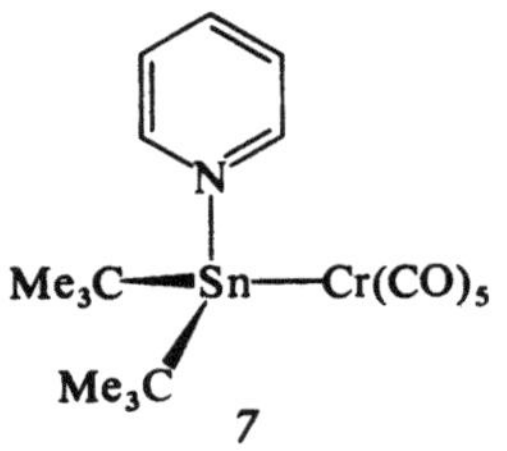

7

When in compounds of type *7* no discrimination can be made between acceptor and donor, one is confronted with molecules which may still possess tin atoms of the formal oxidation state +2 but which are structurally more related to tin(IV) compounds. Classical examples are polystannylenes $(SnR_2)_n$ which may form six-membered rings as in $(Ph_2Sn)_6$ (*8*) [34] (see also Chapter 4).

The square-pyramidal arrangement may be illustrated by two chemically very different compounds: SnO (9) [3] and phthalocyaninetin(II) (*10*) [35]:

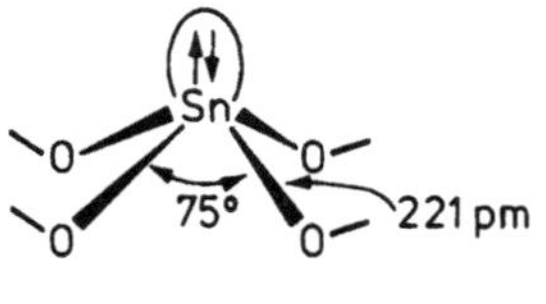

c.f. of *9*

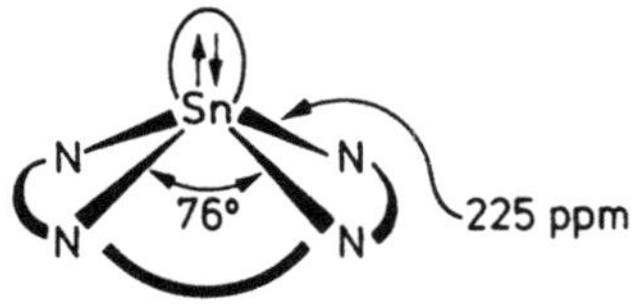

c.f. of *10*

The resemblance of the coordination polyhedra in *9* and *10* is really amazing. The same is true for two Ψ-trigonal bipyramidal arrangements as a comparison of the ionic compound α-$SnWO_4$ (*11*) [36] and the chemically completely different molecular complex Sn(—OC(Ph)CHC(Me)=O)$_2$ (*12*) [37], which contains two chelating acetylacetonato ligands, reveals.

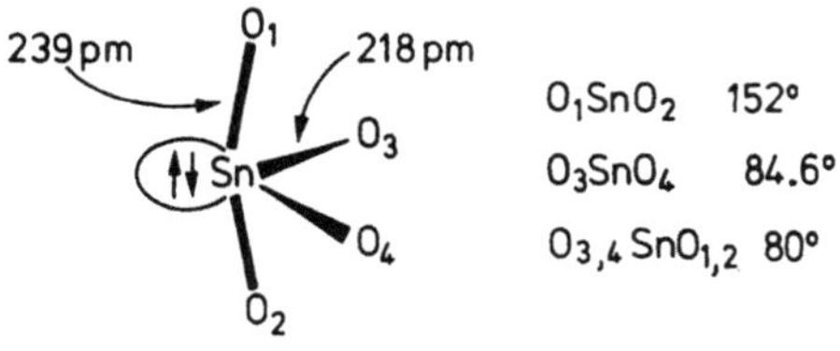

c.f. of *11*

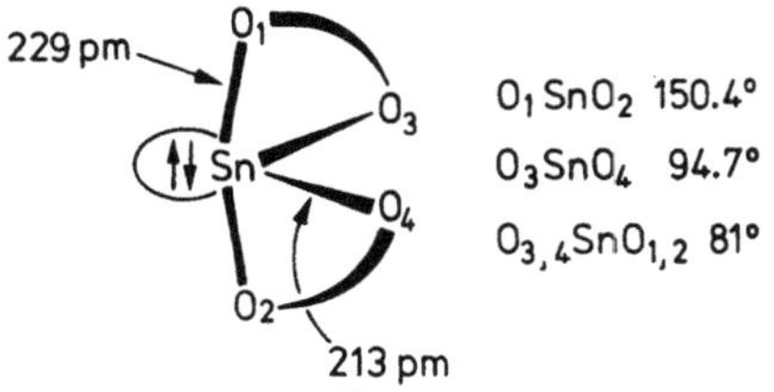

c.f. of *12*

It may be noted that, as expected, the equatorial distances are shorter than the axial. There is, of course, a close relationship between the square-pyramidal and the Ψ-trigonalbipyramidal coordination sphere which are often difficult to discriminate when the tin-ligand bond distances in equatorial and axial positions do not differ significantly.

Some examples of higher coordination numbers of tin are cited in Tables 8–11. Since they are of minor importance in molecular tin(II) chemistry (except [77,78]) they are not discussed here in detail (for a further description see also [3–5]).

3.2.2 Correlation between Tin-Ligand Distances and Coordination Numbers

In Section 3.2.1 we have deduced the dependence of the geometries of tin-ligand arrangements on the coordination number. We will now study the distances between the tin(II) atom and the ligand when the coordination number is changed. In Tables 8–11 the mean distances between the central tin atom and four different "ligands" (N, O, F and S) are listed. As expected, these distances increase with rising coordination number which is graphically illustrated in Fig. 2. Molecular as well as ionic compounds, with the same atom in α-position to the tin atom, show similar tin-ligand distances.

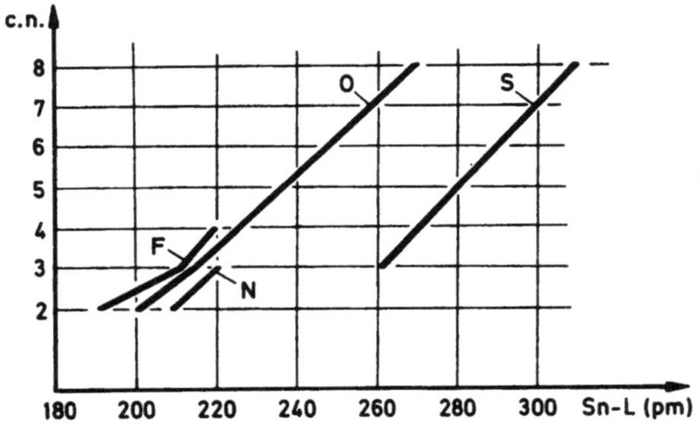

Fig. 2. Correlation between bond lengths and coordination numbers (c.n.). A mean value for each c.n. has been calculated from Tables 8–11; single value (except for c.n. = 2) have been omitted

The following points are important:

1) Tables 8–11 and Fig. 2 may be used to evaluate the coordination number in any tin(II) compound. It is well known that the correct coordination sphere of the structures of solid tin compounds is often difficult to determine. Not in every compound can a distinct difference in the tin-ligand distances around the central atom separating the first coordination sphere from the second one be found. Most of the compounds compiled in the tables have been selected with respect to a deviation of about 30% from the mean distance, according to the first coordination sphere. From Table 12 it follows that distances which are at the extremes of one coordination number (with respect to the mean value) can often be better reclassified by a different coordination number. For example, $SnSO_4$, a compound which is often described by a tin atom surrounded only by three oxygen atoms [58] can be better represented by a coordination of 12 on the tin atom, according to Table 9. Another example is the recently synthesized $K_2Sn_2O_3$ containing two tin atoms, each surrounded by three very narrow oxygen atoms (Sn(1)—O = 196 pm; Sn(2)—O = 209 pm) [79]: it follows from Table 9 that this compound cannot be classified in the usual way; the K cations seem to influence the Sn—O bonds considerably;

2) Taking the curves of oxygen, sulfur and fluorine in Fig. 2 as a reference it seems that a linear correlation can be established between the c.n. and the bond distances (c.n. running from 3 to 8). A rise of the c.n. by one unit increases the mean Sn-ligand distance by about 10 pm;

3) The stannylenes SnF_2, $Sn(OX)_2$ and $Sn(NX_2)_2$ have a remarkably smaller Sn-ligand distance than that resulting from a linear extrapolation of higher coordination numbers (crack in the linear curves of Fig. 2). This effect decreases in the order F > O > N. The differences in the tin-ligand distances between c.n. 2 and 3 are for F: $\Delta = 20$ pm, for O: $\Delta = 15$ pm, and for N: $\Delta = 13$ pm. As the electronegativities decrease in the same order, F > O > N, it seems reasonable to explain this phenomenon by extra ionic components in the bonding of stannylenes.

Table 12. Synthetic routes to molecular tin(II) compounds

X_2Sn	Method according to Scheme 1	Reactant	Ref.
$(C_5H_5)_2Sn$, *13*	a	$2\,XNa + SnCl_2$	82)
$(Me_5C_5)_2Sn$, *3*	a	$2\,XLi + SnCl_2$	83)
$[(Me_3Si)_2CH]_2Sn$, *14*	a	$2\,XLi + SnCl_2$	84)
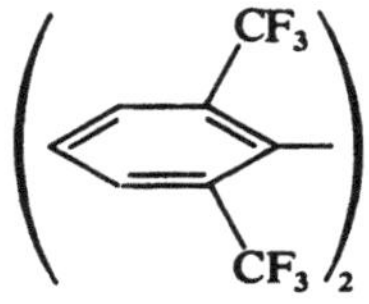 Sn, *15*	a	$2\,XLi + SnCl_2$	85,8)
$[(Me_3Si)_2N]_2Sn$, *16*	a	$2\,XLi + SnCl_2$	86,87)
$Me_2Si(Me_3CN)_2Sn$, *1*	a	$XLi_2 + SnCl_2$	88)
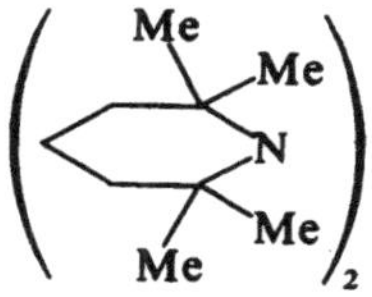 Sn, *17*	a	$2\,XLi + SnCl_2$	89)
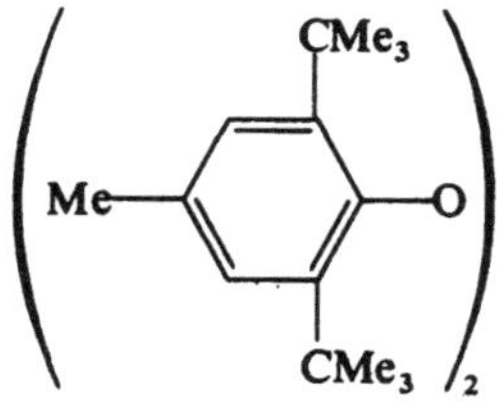 Sn, *2*	a	$2\,XLi + SnCl_2$	28)
$[(Me_3C)_2P]_2Sn$, *18*	a	$2\,XK + SnCl_2$	90)
$(Et_2NCS_2)_2Sn$, *19*	a	$2\,XK + SnCl_2$	91)
$(Me_3C)_2As(Cl)\,Sn$, *20*	a	$XSiMe_3 + SnCl_2$	92)
$(C_5H_5)\,ClSn$, *21*	b	$^1/_2\,(C_5H_5)_2Sn + {}^1/_2\,SnCl_2$	93,94)
$[(Me_3Si)_2N]\,ClSn$, *22*	b	$^1/_2\,[(Me_3Si)_2N]_2Sn + {}^1/_2\,SnCl_2$	86)
$(RC(O)\,CHCR'O)_2Sn$	c	$(MeC_5H_4)_2Sn + 2\,RC(O)\,CH_2C(O)\,R'$	95,96,97)
$(RCOO)_2Sn$	c	$(MeO)_2Sn + 2\,HOOCR$	98)
$RN(CH_2CH_2O)_2Sn$	c	$(MeO)_2Sn + (HOCH_2CH_2)_2NR$	99)
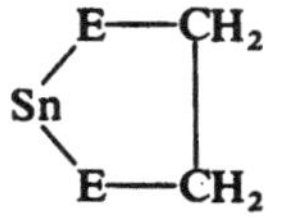	c	$(MeO)_2Sn + HE{-}CH_2CH_2{-}EH$ (E = O, S)	100)
$(Me_2Si)_3(NMe)_5Sn_2$, *23*	c	$3\,Me_2Si(Me_3CN)_2Sn + 3\,(HMeN)_2SiMe_2$	101)

Table 12 (continued)

X_2Sn	Method according to Scheme 1	Reactant	Ref.
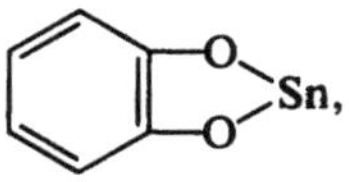 *24*	d	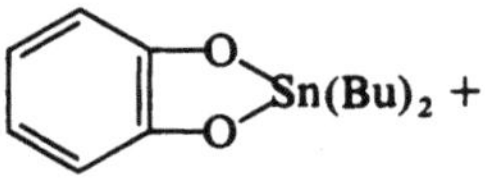 + $SnCl_2$	102)
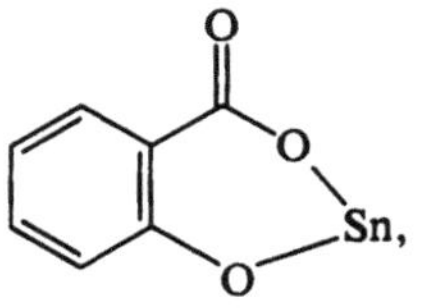 *25*	d	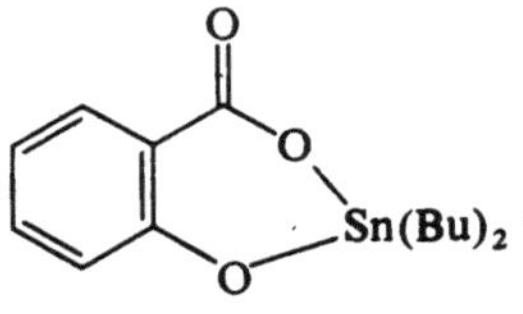 + $SnCl_2$	102)
$MeB(NSiMe_3)_2Sn$, *26*	e	2 $LiN(SiMe_3)$ BMe_2 + $SnCl_2$	39)
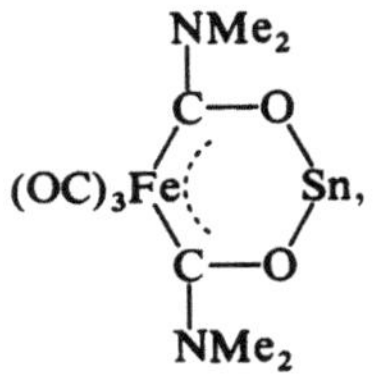 27	e	$^1/_2$ $[(Me_2N)_2Sn]_2$ + $Fe(CO)_5$	103)

Although the results obtained by comparison of bond lengths and coordination numbers are illustrative they should be used with care. Since this approach does not take into account the nature of the ligands it is only a very rough one. It is nevertheless remarkable, that solid and molecular tin(II) compounds (which differ also chemically) possess similar geometries and distances around the tin atom.

4 Syntheses and Structures of Molecular Tin(II) Compounds

4.1 Synthetic Routes to Molecular Tin(II) Compounds

The classical route to synthesize organic tin(II) compounds is the thermolysis or photolysis of tetravalent tin compounds. The driving force of these reactions may be either weak tin-tin bonds or the formation of stable compounds besides the desired stannylenes which all polymerize. Some examples taken from the latest studies of W. P. Neumann and coworkers [7,80,81] are illustrated by Eqs. (1–3).

$$Bu_3Sn - SnBu_2Cl \xrightarrow{\Delta T} Bu_3SnCl + [Bu_2Sn]_x \tag{1}$$

$$(Bu_2Sn)_\infty \xrightarrow{h\nu} \{Bu_2Sn\} \longrightarrow [Bu_2Sn]_x \tag{2}$$

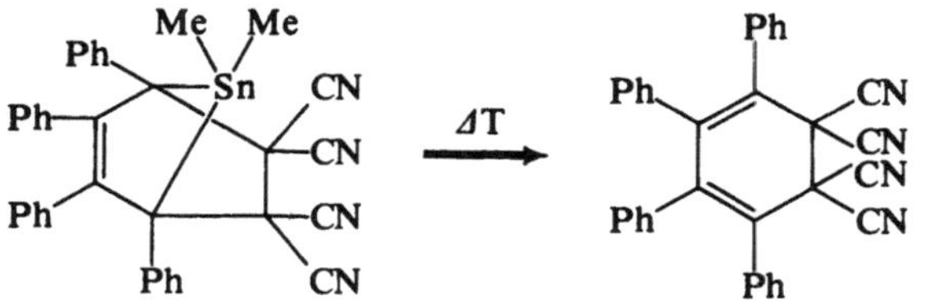

$+ \{SnMe_2\}$ (3)

Until now, none of these reactions has ever lead to stable monomolecular stannylenes. Nevertheless, trapping reactions have confirmed the presence of intermediate Sn(II) species [7,80,81].

Stable stannylenes can be synthesized via routes which are combined in Scheme 1 whereas typical examples for these reactions are listed in Table 12.

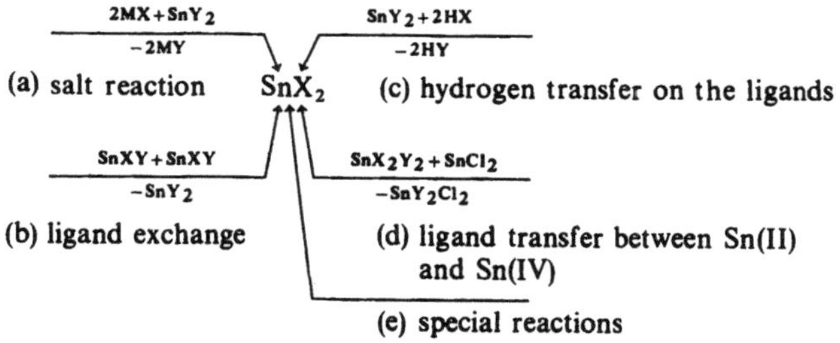

Scheme 1

By far the most important reaction is the salt elimination reaction (a). Most of the known monomeric stannylenes have been synthesized via this route which is unaffected by the bulkiness of the substituent. In all cases, tin(II) chloride is taken as the tin component, because it dissolves quite well in ethers. The yields of the stannylenes are relatively high and may attain 95%.

Process (b), which should normally be written

$$SnX_2 + SnY_2 \rightarrow 2\,SnXY\,, \tag{4}$$

is important in tincyclopentadienyl chemistry as well as in other special cases [86,94,98]. The ligand exchange can be considered as an equilibrium reaction which is shifted to SnXY because of the low solubility of the compound, SnY_2 corresponding to $SnCl_2$ in nearly all cases.

The hydrogen-transfer reaction (c) from one ligand X to the other ligand Y is often utilized when chelating ligands have to be bonded to tin. The leaving group HY may be cyclopentadiene, methanol, or a simple amine such as dimethylamine. The entering ligand is often more acidic. The coordination sphere at the tin atom is at least 3, but may also be 4 or higher.

Ligand transfer from tin(IV) to tin(II) compounds with appropriate ligands such as chlorine (d) has been used infrequently in the synthesis of stannylenes. The only reactions reported so far have been performed with oxygen as substituents at tin (see also Sect. 6.4.3).

Under "special reactions" (e) two procedures are combined which are unique and can be used for further preparations.

In the synthesis of the boron-containing four-membered cycle *26* the reaction is started by a salt elimination of type (a) followed by an intramolecular condensation

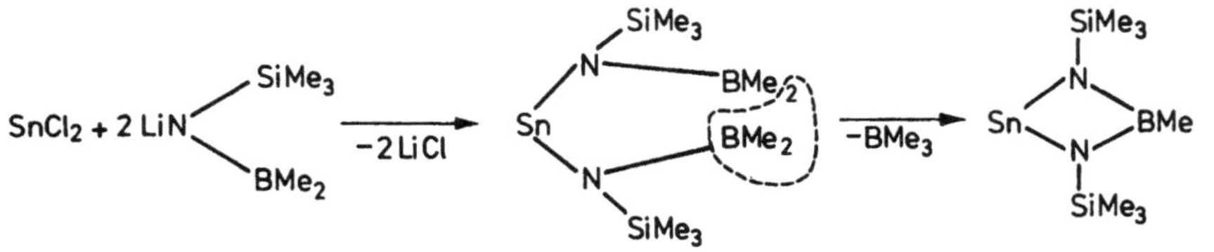

(5)

26

with elimination of trimethylboron [39]. In the second example of Table 12 an addition of the Sn—N bond to the carbonyl groups is thought to result in the formation of the six-membered ring *27* which is believed to be stabilized by mesomerism [102].

4.2 Structures of Molecular Tin(II) Compounds

As previously pointed out in Chapter 2, monomeric stannylene can be in equilibrium with oligomeric species which are formed by tin-tin or tin-substituent intermolecular interactions. The tendency for the formation of the oligomers increases the more the molecules approach one another. Thus, when passing from the vapor to the liquid phase and finally to the solid state, the molecules usually exhibit quite different structures. In Table 13 examples of the corresponding structural changes are given.

Before studying some examples more closely, let us consider some cases which are not listed in Table 13. There are numerous compounds SnX_2 which are definitely monomeric but are nevertheless no carbene analogs since their valence electron number at the tin atom is at least eight. These compounds contain chelating ligands which can stabilize the carbenoid tin atom due to intramolecular Lewis acid-base interactions as shown by structure A and B (see also Chapter 3).

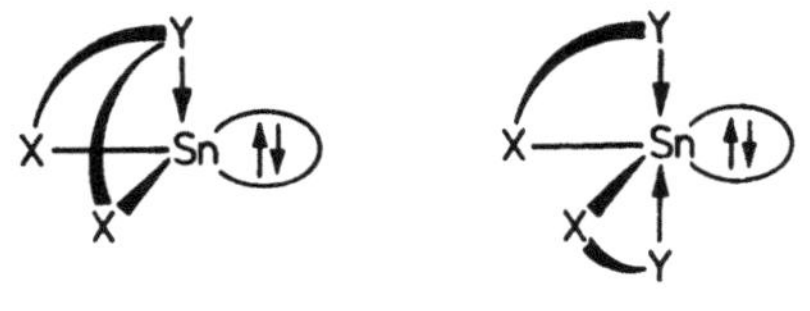

A B

Table 13. Dependence of the structure of monomeric tin(II) compounds on the substituents and the phase

Compound	Vapor	Solution or melt	Solid phase	Ref.
$(C_6H_5)_2Sn$, *8*	—	polymeric	polymeric and cyclic	34)
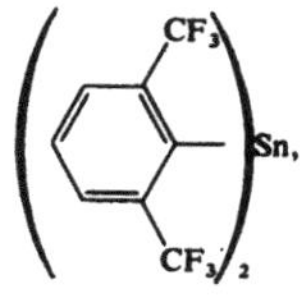 *15*	—	monomeric	unknown	8,85)
$(C_5H_5)_2Sn$, *13*	monomeric	monomeric, polymerizes on standing	polymeric	104,105,106)
$(C_5H_5)ClSn$, *21*	monomeric	monomeric only in coordinat. solvents	polymeric, one-dimensional	93,94,107)
Cl_2Sn, *28*	monomeric	ionized or strongly coordinated	polymeric	20)
$(Me_5C_5)_2Sn$, *3*	monomeric	monomeric	monomeric	29,83)
$[(Me_3Si)_2CH]_2Sn$, *14*	monomeric	monomeric	discrete dimers	21,108)
$[(Me_3Si)_2N]_2Sn$, *16*	monomeric	monomeric	at least dimeric	38,109)
$[(Me_3Si)_2N]ClSn$, *22*	—	polymeric	polymeric	86)
$(Me_2N)_2Sn$, *29*	monomeric	dimeric	unknown	110)
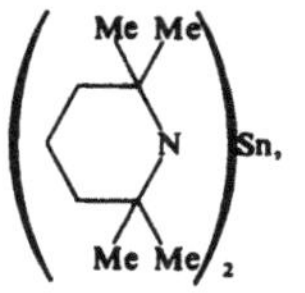 *17*	monomeric	monomeric	unknown[a]	89)
$[(Me_3C)_2N]_2Sn$, *30*	—	monomeric	unknown	89)
$[(Me_3C)_2P]_2Sn$, *18*	—	dimeric	unknown	90)
$[(Me_3C)_2P]ClSn$, *31*	—	polymeric	unknown	111)
$[(Me_3C)_2As]ClSn$, *20*	—	polymeric	unknown	92)
$Me_2Si(Me_2HCN)_2Sn$, *32*	monomeric	dimeric	dimeric	112)
$Me_2Si(Me_3CN)_2Sn$, *1*	monomeric	monomeric	monomeric[b]	27,88,112)
$(Me_3C)_2Sn(Me_3CN)_2Sn$, *33*	—	monomeric	unknown	113)
$MeB(Me_3SiN)_2Sn$, *26*	—	dimeric	dimeric	39)
$(CH_2)_x(Me_3SiN)_2Sn$, *34a, b, c*, x = 2, 3, 4	monomeric	dimeric or polymeric	unknown	114)

Table 13 (continued)

Compound	Vapor	Solution or melt	Solid phase	Ref.
$(MeO)_2Sn$, *35*	at least dimeric	polymeric or coordinated	polymeric	[115,116]
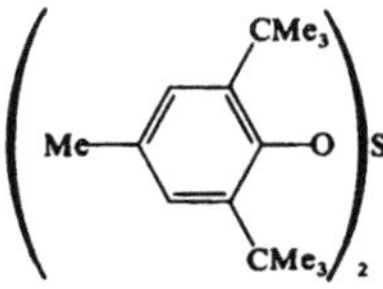 , *2*	monomeric	monomeric	monomeric	[28]
$(OC)_3Fe(\cdots C(NMe_2){-}O{-})_2Sn$, *27*	–	monomeric	unknown	[103]

ᵃ the corresponding Ge compound has a monomeric structure in the solid phase
ᵇ two solid modifications are known: in the triclinic phase only dimers are present while in the monoclinic phase four dimeric and four monomeric units are present simultaneously

Atoms X and Y can either be different of identical. While for type B several examples are known (e.g. bis-β-oxoenolates [95–97], bis-carboxylates [98], bis-sulfonates [98], bis-dithiocarbonates [91]), only few compounds of type A have been synthesized and characterized up to now [99,117].

While diphenyltin (*8*) is polymeric (due to its crystal structure is can be isolated as a six-membered ring compound, Fig. 3), the corresponding compound *15* is monomeric in solution [85]. The effect of the two CF_3-groups in ortho-position is believed to be predominantly of steric nature. Dicyclopentadienyltin(II) was the first organometallic tin compound whose structure has been established to be monomeric and bent in the vapor [104–106]. It polymerizes slowly in solution and presumably exhibits a two-dimensional polymeric structure as a solid [106]. If one cyclopentadienyl ligand, which is pseudo π-bound to the central atom, is replaced by chlorine, the resulting compound is not longer soluble in non-polar solvents but only in coordinating solvents such as tetrahydrofurane [94]. Solid $(C_5H_5)ClSn$ is also polymeric [107], and the molecules

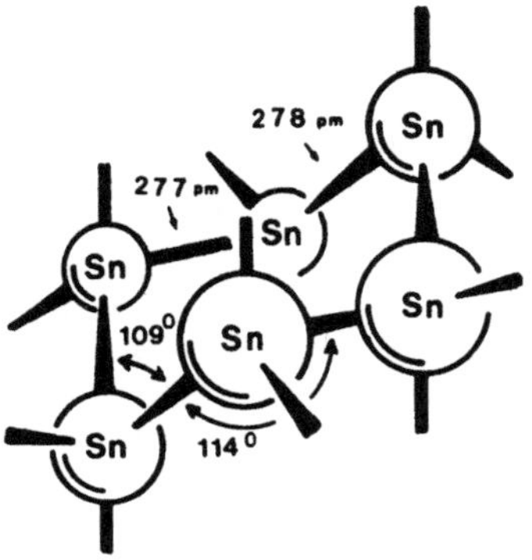

Fig. 3. Sn_6-skeleton of the ring compound $(SnPh_2)_6$ (*8*) [34]

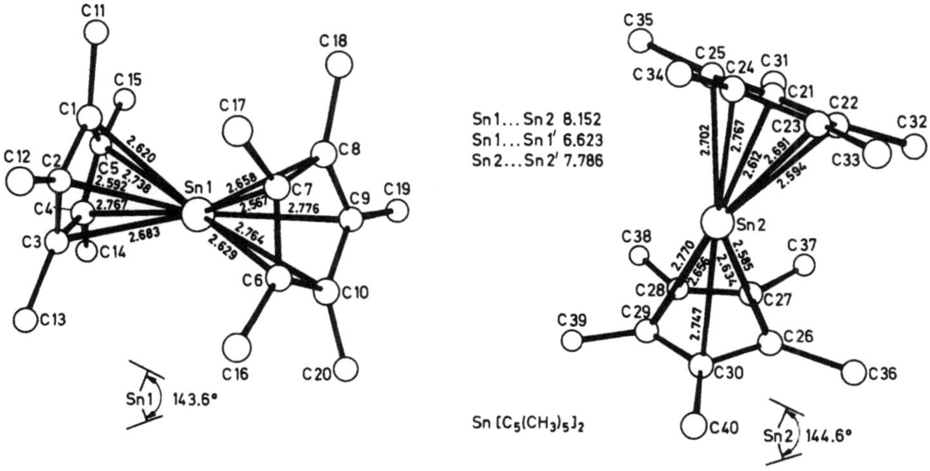

Fig. 4. Crystal structure of bis(pentamethylcyclopentadienyl)-tin(II) (*3*). Reprinted with permission from Chem. Ber. *113*, 760 (1980). Copyright by Verlag Chemie

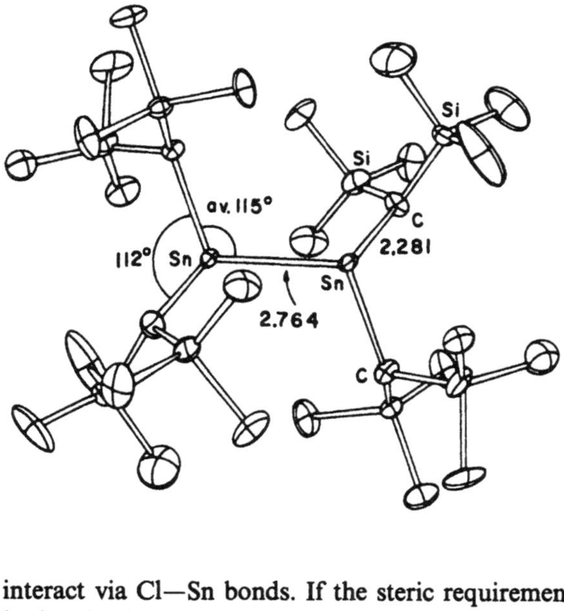

Fig. 5. Crystal structure of bis[di-(trimethylsilyl)methyl]tin(II) (*14*). Reprinted with permission from J. Chem. Soc., Chem. Commun. *1976*, 261. Copyright by The Chemical Society

interact via Cl—Sn bonds. If the steric requirement of the cyclopentadienyl groups is drastically raised by substitution of the hydrogen atoms by methyl groups the resulting compound, $(Me_5C_5)_2Sn$, is found to be monomeric in all phases [83] as demonstrated in Fig. 4.

The structure of the organometallic tin(II) compound *14*, which was the first stable bivalent tin compound in non-polar organic solvents [84], is changed when passing from the solution to the solid state [21, 108]. In the crystal discrete dimers are present (Fig. 5). Since the tin atom is pyramidal and the Sn—Sn distance quite large (276 pm), no normal σ, π-double bond can be responsible for this geometry.

M. F. Lappert has therefore proposed a double Lewis acid-base interaction of type C [108, 118].

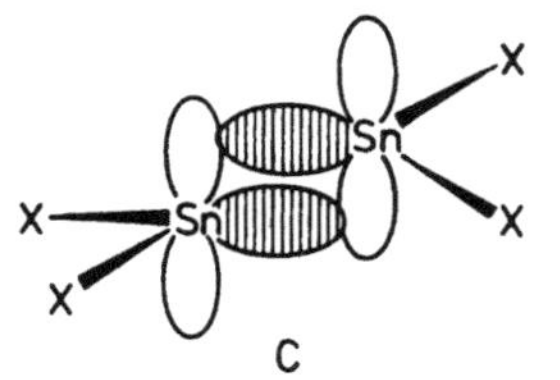

C

The substituent $CH(SiMe_3)_2$ is very suitable for the stabilization of tin(II) compounds: in addition to its large steric requirements it should be a strong σ-donor, trimethylsilyl groups being located in β-position to the tin atom. It is not astonishing that the isoelectronic bis(trimethylamino) group stabilizes monomeric stannylenes as well, but nothing is known about the structure of the solid *16*. In the vapor phase the molecule is bent at the tin atom by an angle of 96°, the nitrogen atoms being strictly trigonal planar and the whole molecule adopting C_{2v}-symmetry (Fig. 6) [38]. Again, substitution of one bis(trimethylsilyl)amino group of *16* by chlorine leads to polymerization (molecule *22*).

A very interesting sequence of compounds is listed in Table 13 beginning with compound *29* up to *20*. While bis(dimethylamine)tin(II) is dimeric in solution, the molecule with the much more bulky substituent *17* is monomeric. In addition, substitution of the methyl groups by tert-butyl groups in *29* reduces aggregation (compound *30*). The corresponding phosphorus compound *18* is again dimeric, indicating the stronger Lewis-base property and the larger covalent radius of phosphorus. Substitution of one bis(tert-butyl)phosphinyl and bis(tert-butyl)arsinyl group by chlorine leads to polymeric *31* and *20*.

Some cyclic nitrogen-containing stannylenes are listed in Table 13. While the 1,3-diisopropyl-1,3-diazastannetidine *32* is dimeric in solution and in the solid state, the corresponding 1,3-di-tert-butyl derivative is strictly monomeric in the vapor phase and in solution. Two solid phases can be prepared: when a pentane solution of *1* is allowed to crystallize, a monoclinic phase forms with a 1:1 mixture of monomers and dimers. If the melt is crystallized, only dimeric units are found in the triclinic crystal lattice (see Figs. 7 and 8). These experiments clearly show that the degree of aggregation of the molecules directly depends on the degree of "dilution" of the compounds in the "solvents".

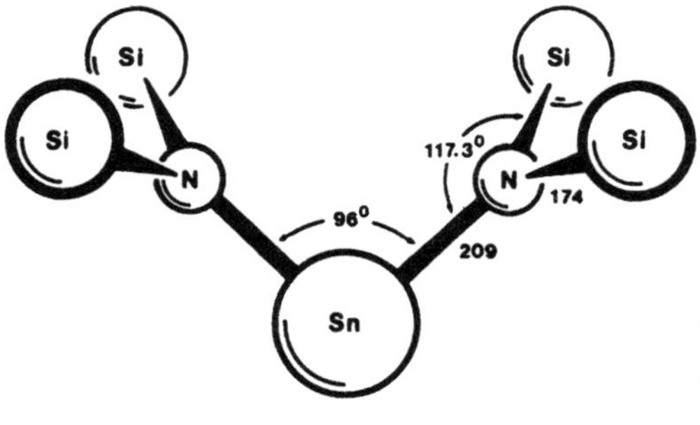

Fig. 6. Structure of $[(Me_3Si)_2N]_2Sn$ (*16*) in the vapor [38]

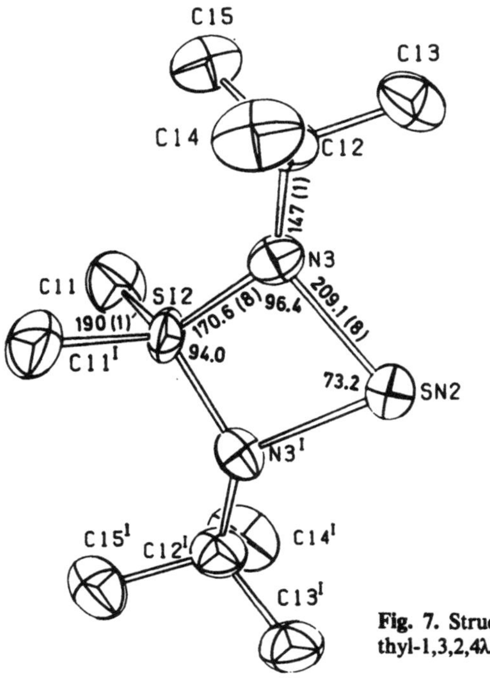

Fig. 7. Structure of monomeric 1,3-di-tert-butyl-2,2-dimethyl-1,3,2,4λ^2-diazasilastannetidine (*1*) in the crystal [27]

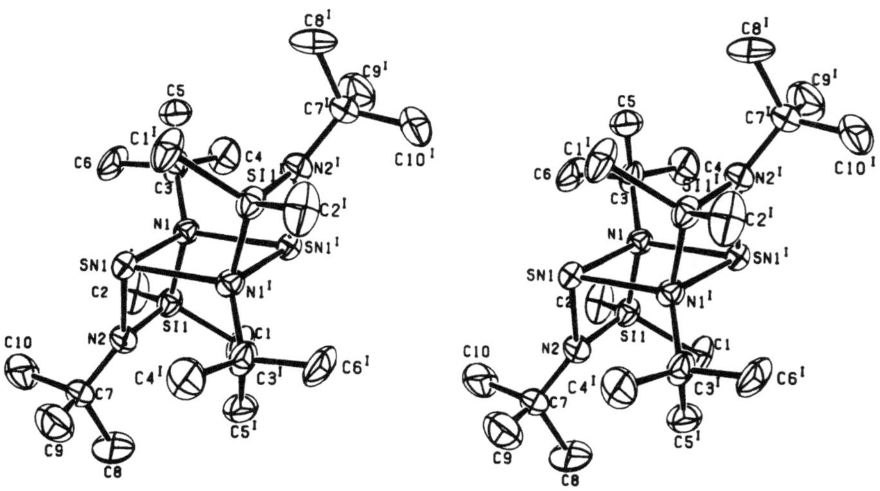

Fig. 8. A stereographic view of the corresponding dimeric unit of Fig. 7. Reprinted with permission from Z. Naturforsch. *33b*, 11 (1978). Copyright by Verlag Zeitschr. für Naturforsch.

Substitution of the dimethylsilyl group by bis(tert-butyl)-stannyl does not change the structure in solution, e.g. *33* is found to be monomeric. A very interesting dimer is *26*. In contrast to the centrosymmetrical dimer of *1* (C_i-Symmetry), *26* has a twofold axis (C_2, see Fig. 9). This special structure may be due to intramolecular Lewis acid-base interactions between the boron and nitrogen atoms [39]. Nevertheless,

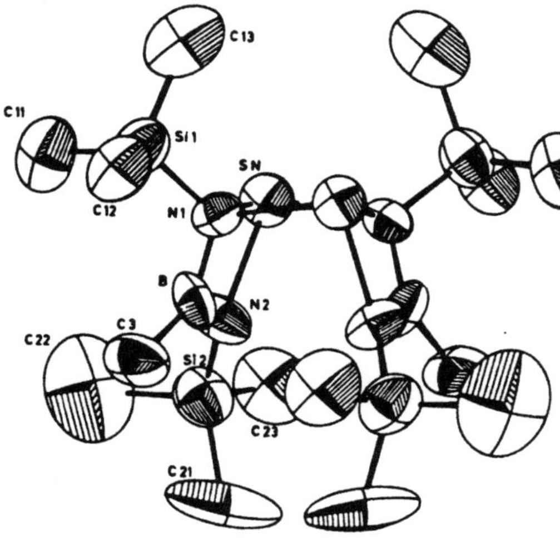

Fig. 9. Crystal structure of dimeric $MeB(Me_3SiN)_2Sn$ (*26*). Reprinted with permission from Chem. Ber. *112*, 3677 (1979). Copyright by Verlag Chemie

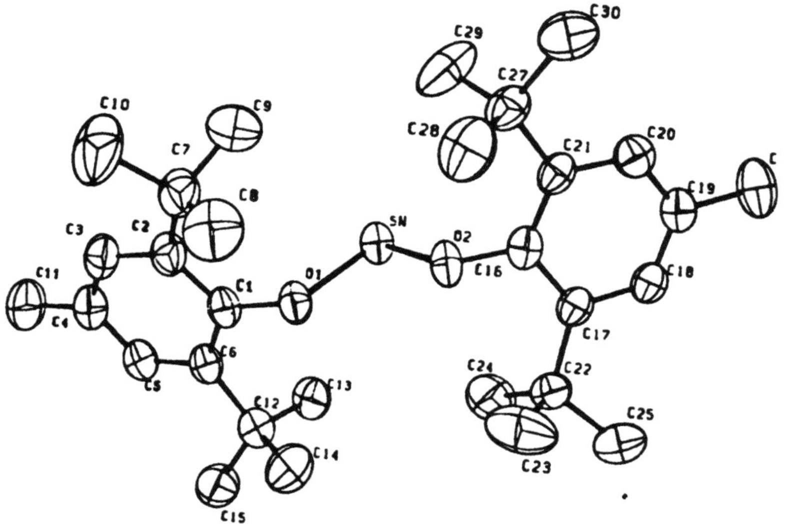

Fig. 10. Crystal structure of

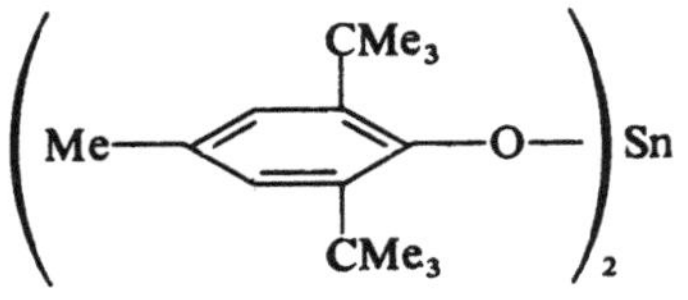

(*2*). Reprinted with permission from J. Amer. Chem. Soc. *102*, 2088 (1980). Copyright by the American Chemical Society

it demonstrates very well that the trimethylsilyl group at the α-nitrogen atoms is not so efficient in stabilizing monomeric stannylenes as the tert-butyl ligand. This can also be deduced from the structures of the compounds *34a*, *b*, *c*.

There are numerous examples of tin(II)-oxygen compounds which demonstrate the high aggregation tendency of this type of compounds [119, 120]. A typical representative is dimethoxystannylene (Table 13) which is oligomeric in all three phases. Only very recently have M. F. Lappert and coworkers shown that oxygen-containing ligands can also be considered as substituents in strictly divalent stannylenes. Besides *1* and *3*, *2* is the only known example of a stannylene which is monomeric in the vapor phase, in the melt and as a solid (Fig. 10). Compound *27* is the first cyclic stannylene bridged by oxygen atoms and thus not coordinated in non-polar solvents. Unfortunately, no direct structural proof has been given for this molecule at the moment.

5 Conclusions from Chapters 3 and 4

Apart from the determination of the structures of stannylenes by diffraction methods (X-ray or electron diffraction) many other physico-chemical techniques can be exployed to characterize these compounds more completely. Besides the classical methods such as IR-, Raman-, PE-, UV- and NMR-spectroscopy, Mößbauer-119 m-tin spectroscopy is widely used for the determination of the oxidation states of tin atoms and of their coordination [100,102,114,118,120–123]. It is not in the scope of this report to study the dependence of Mößbauer constants such as isomer shift and quadrupole splitting on structural parameters. Instead, we want to concentrate on one question: Which information can we deduce from the structure of stannylenes to evaluate their reactivity?

The following points are of general relevance:

1) Molecular compounds of bivalent tin are in all cases bent molecules, the angle between the substituents approaching 90°. Only the π-pseudo-bound cyclopentadienyl derivatives deviate considerably from this angle (120 and 144°, respectively);

2) Bulky substituents are necessary for the stabilization of stannylenes which may have different structures in solution and in the solid state. The substituents envelope the tin atom from the back side, thus restricting its reactivity. Reactions should therefore occur predominantly from the front side;

3) The bond lengths between tin(II) and atoms with lone electron pairs like nitrogen or oxygen are quite small (200 and 209 pm compared with 208 and 215 pm as expected (see Chapter 3)) suggesting intramolecular electron compensation of the π-type:

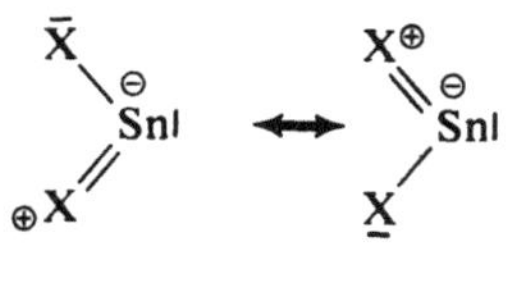

This interaction cannot be very important, as may easily be deduced comparison of compounds *1* and *16* (see also Figs. 6 and 7): Whereas in *1* the two filled π-orbitals of the sp^2-hybridized nitrogen atoms are equiplanar to the p_z-orbital of the tin atom, they are orthogonal in molecule *16*, the bond distances being rigorously equal;

4) Stannylenes are in the first place Lewis acids (electron acceptors) as can be easily derived from the structures of the solids (Chapter 3). When no Lewis bases (electron donors) are present, they may also act as Lewis bases via their non-bonding electron pair (see polymerization of organic stannylenes).

A quantitative evaluation of the acid properties is quite difficult. One example is nevertheless quite instructive: in 1,3-di-tert-butyl-2,2-dimethyl-1,3,2,4λ^2-diazasilastannetidine (*1*) the tin atom can be replaced by the much smaller aluminium-methyl group [124]. While *1* is monomeric in benzene, *36* is dimeric:

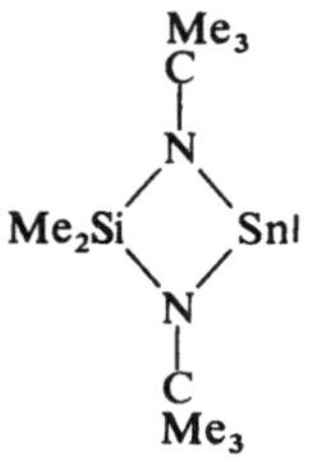

1

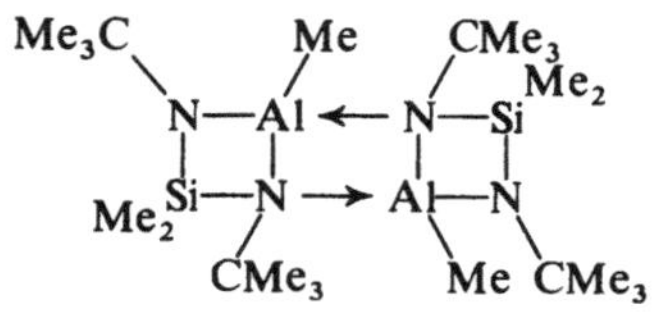

36

5) When stannylenes are allowed to react with nucleophiles Y| the attack of the electron donor proceeds stereospecifically and orthogonal to the SnXX′-plane. This can be demonstrated by the Lewis base-adducts of the stannylenes which exclusively exhibit structures of type A or B, depending on the number of nucleophiles present.

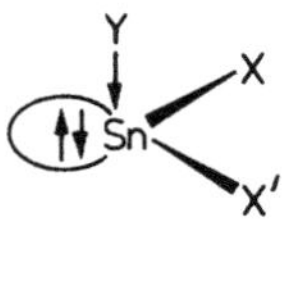

A

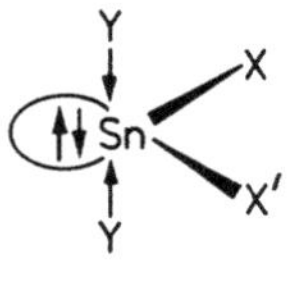

B

If X ≠ X′ ≠ Y compounds of structure A have a center of chirality and the R- and S-enantiomers should be optically active. Since in Lewis acid-base reactions exchange equilibria are often expected to be formed via transition state B, it seems quite difficult to synthesize one pure enantiomeric form;

6) Stannylenes should be easily oxidized to molecular tin(IV) compounds as reflected by their low ionization potentials. In alkyl-substituted stannylenes the ionization energy is 7.42 eV while in stannyleneamides it is 8.38 eV [125]. In any case, the tin atom is sterically easily accessible by reagents (see also point 2);

7) While the acidic behavior of the tin atom in stannylenes can be foreseen incontestably, its Lewis-base properties are much more difficult to evaluate. Taking the angle X_2Sn as a reference, the basicity should increase at angles larger than 90° as in the case of the cyclopentadienyltin, compounds where the orbital of the lone electron pair acquires more p- than s-character [8]. This picture is surely too simple: with few exceptions all tin(II) compounds show a high degree of stereochemical activity of the lone electron pair whether the ligands are bound nearly orthogonal to one another or not.

6 Reactions of Stannylenes SnX_2

In contrast to carbenes the singlet electron configuration in stannylenes SnX_2 is much more stable; this implies that the non-bonding electron pair can remain unchanged during a reaction. Consequently, this reaction center and other centers must be considered in a reaction pathway multiplying the reaction possibilities compared with the isoelectronic carbenes.

Following the ideas exposed in Chapter 4 different reactive centers can be distinguished in stannylenes:

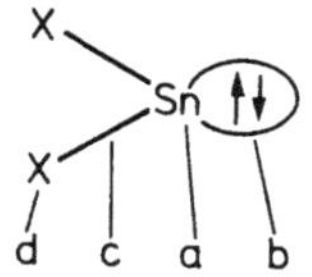

(a) the low-valent unsaturated tin atom,
(b) the non-bonding electron pair,
(c) the heteropolar Sn—X bond, and
(d) the Lewis-base properties of substituents X.

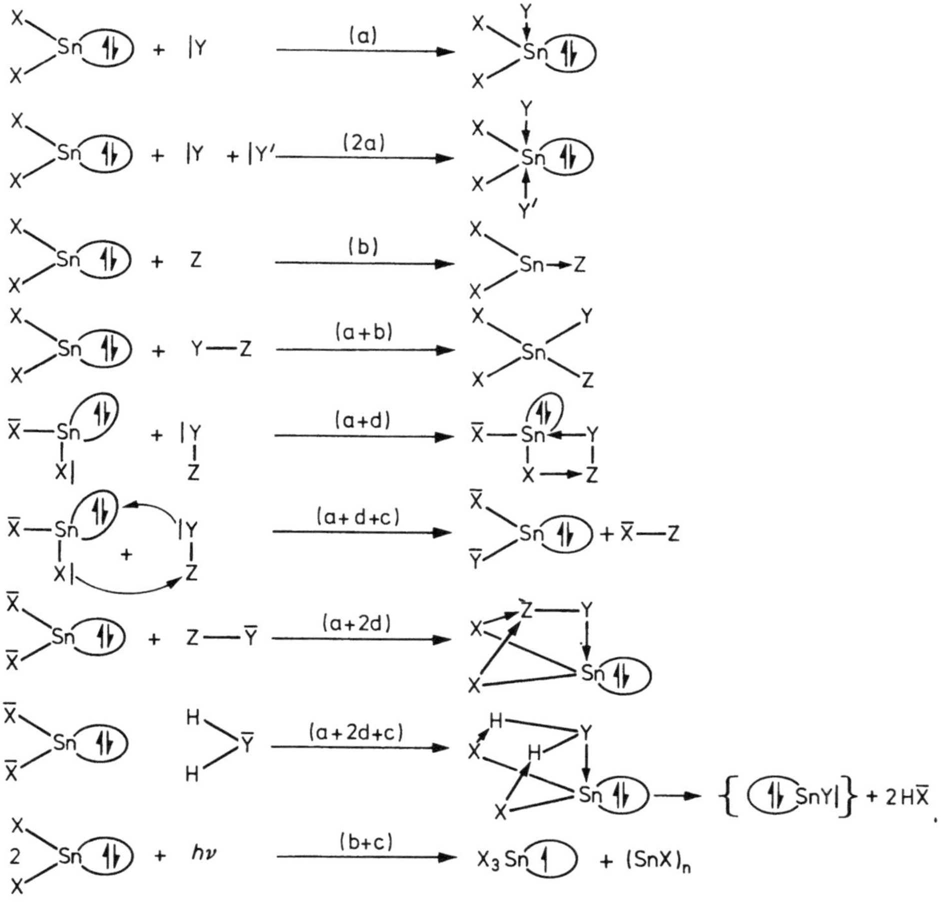

Scheme 2

Which of the different centers (a)–(d) will enter in a reaction depends of course on the reactant. Either only one or more centers may be attacked simultaneously. In Scheme 2 several possibilities are compiled, the details being discussed in the following chapters. The letters above the arrows in this scheme refer to the different centers involved in the reaction.

6.1 SnX_2 Reacting as a Lewis Acid

6.1.1 Formation of Simple Adducts

It has been known for a long time that nearly any tin(II) compound can form stable adducts with bases [126]. For example, the halides $SnHal_2$, which do not form stable bivalent tin compounds in solution, can nevertheless be stabilized by the addition of bases:

$$SnHal_2 + B \rightarrow SnHal_2 \cdot B \quad (6)$$

$$SnHal_2 + 2\,B \rightarrow SnHal_2 \cdot 2\,B \quad (7)$$

$$SnHal_2 + Hal^- \rightarrow SnHal_3^- \quad (8)$$

As base B several electron donors can be employed, e.g. amines [127], hydrazines [128], amine oxides [129], ethers [126], phosphanes [130, 131], and even charged ligands such as halogen anions [126] and many others [126]. There may be formed either 1:1 or 1:2 adducts as in Eq. (7), an equilibrium being assumed to be established between these two adducts:

$$SnHal_2 \cdot 2\,B \rightleftarrows SnHal_2 \cdot B + B \quad (9)$$

C. C. Hsu and R. A. Geanangel [132] have studied the properties of these adducts for all tin(II) halides being coordinated with one or two trimethylamine molecules. As in the case of the trimethylamine-borontrihalide adducts the acceptor strength of the halides with respect to the amine decreases in the order: $SnF_2 > SnCl_2 > SnBr_2 > SnJ_2$. On the other hand, the dissociation according to Eq. (9) is highest for SnF_2, the $SnJ_2 \cdot (Me_3N)_2$ adduct being much more stable than $SnF_2 \cdot (Me_3N)_2$.

Whereas adducts of instable stannylenes with bases are stable, easily isolatable compounds, stable stannylenes form adducts which are difficult to handle and to characterize. Taking $[(Me_3Si)_2CH]_2Sn$ (*14*) and $Me_2Si(NCMe_3)_2Sn$ (*1*) as examples [133, 134], it is found that these two compounds can be coordinated with a base like pyridine (and also with other bases, see Ref. [133]). However, the 1:1 adduct is only stable at −30 °C and decomposes at room temperature. The base attached to the stannylene can easily be removed in a vacuum at reduced pressure. When allowed to recondensate, it again forms a complex with stannylene [134]. The equilibrium formulated by Eq. (10) can thus be shifted to the desired side.

$$X_2Sn \cdot B \rightleftarrows X_2Sn + B \quad (10)$$

Taking into consideration the findings with tert-butylamine in Section 6.6.1 it may be assumed that in solution the equilibrium of these reactions (Eq. (10)) is mostly shifted to the left side, involving however a rapid exchange of base molecules within the adduct. The equilibrium may depend on temperature and the molarity of B.

Generally speaking, the Lewis acidity of the tin atom and, as a consequence, the stability of the formed base adduct can easily be deduced from the structure of the stannylene under consideration. If it is stable and monomolecular in non-polar solvent, it will form weak adducts. If it is not, it will form strong and stable adducts.

6.1.2 Displacement Reactions

The displacement of one base coordinated to a stannylene SnX_2 by another base can easily be achieved if the entering molecule is more basic than the leaving one. If an ether solution of *1* is allowed to react with pyridine, the weak ether adduct is converted to a pyridine adduct at −30 °C [112, 134]:

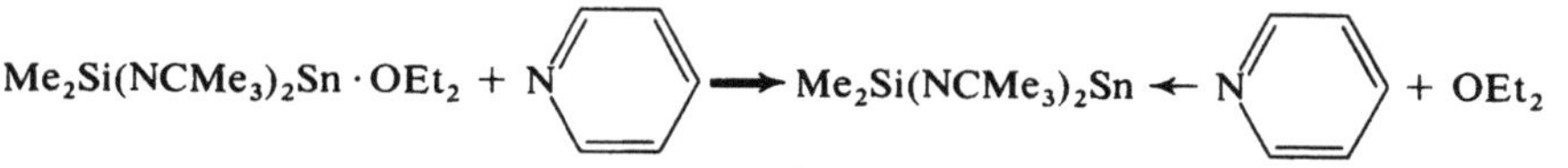

(11)

37

Another illustrative example is described by Eq. (12) [135] which demonstrates that the adduct needs not necessarily contain only a tri-coordinated tin atom.

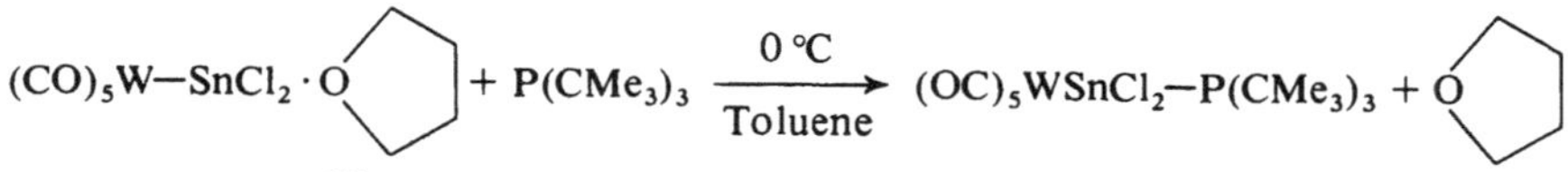

(12)

38 *39*

As in $Cl_2Sn{-}P(CMe_3)_3$ the metal-coordinated *39* may be interpreted as an ylide-type compound. Reactions and physicochemical measurements indicate that the phosphane acts as a simple donor [130].

These displacement experiments can be utilized to evaluate different base properties of ligands which are attached to the tin atom as illustrated by Eqs. (13) [112] and (14) [94].

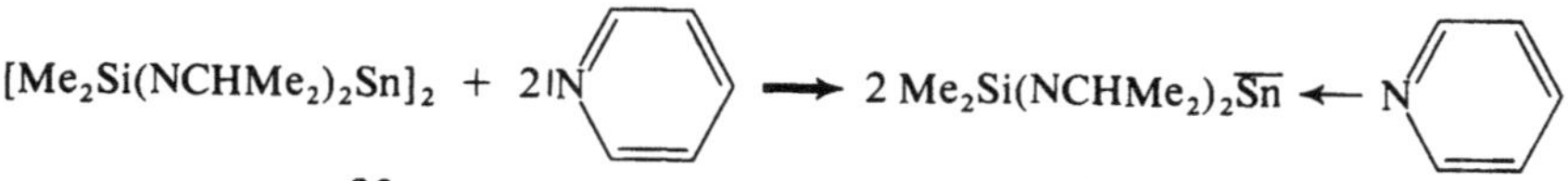

(13)

32 *40*

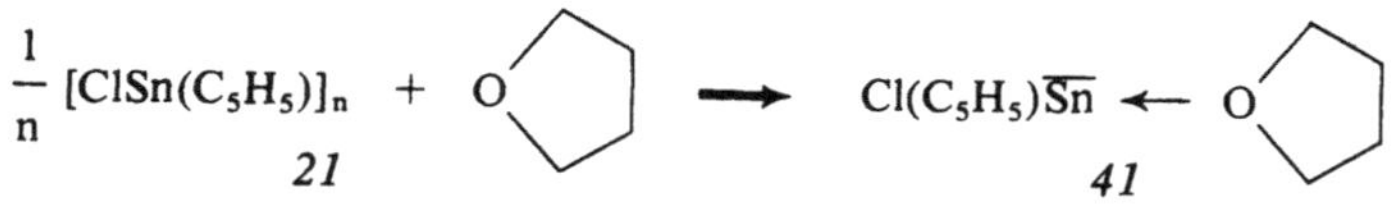

(14)

21 *41*

While in the first case (Eq. (13)) the dimeric unit, which is formed due to nitrogen-tin interactions (Chapter 4), is converted to the simple adduct *40* by the stronger base pyridine, the second case illustrates that the polymer *21* can be dissolved in a coordinating solvent.

6.2 SnX_2 Reacting as a Lewis Base

6.2.1 Formation of Adducts with Main-Group Acids

In 1970 Harrison and Zuckerman [136] reported for the first time the existence of a stable adduct of a stannylene (dicyclopentadienyltin(II)) with a Lewis acid (trifluoroborane). In the meantime, P. G. Harrison was able to demonstrate that these Lewis-acid base interactions can be extended to a variety of main group acids [137, 138] (Eqs. (15) and (16)).

$$2\,(H_5C_5)_2Sn + Al_2X_6 \rightarrow 2\,(H_5C_5)_2Sn - AlX_3 \qquad (15)$$

$X = Cl, Br$ *42a, b*

$$(C_5H_5)_2Sn + Et_2O \cdot BX_3 \rightarrow (C_5H_5)_2Sn - BX_3 + Et_2O \qquad (16)$$

$X = F, Br$ *43a, b*

With BCl_3-etherate ligand-exchange reactions occur, resulting in tin(II) chloride as the main product. The structures of the adducts *42* and *43* have been established mainly on the basis of IR- and Mößbauer data; unfortunately, no direct structural determination has been performed. Mostly from IR data it is believed that the cyclopentadienyl ligands are centrally σ(pseudo π-) bonded the angular geometry

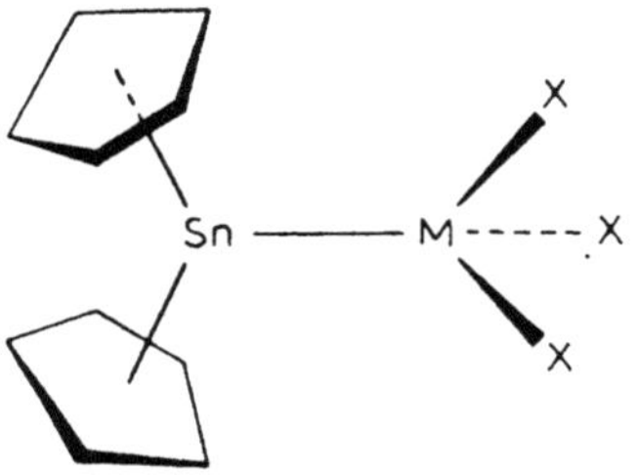

Fig. 11. Proposed structure for $(C_5H_5)_2Sn{-}MX_3$ complexes (M = B, Al). Reprinted with permission from J. Organomet. Chem. *108*, 38 (1976). Copyright by Elsevier Sequoia S.A.

of the free stannylene being retained and MX_3 being coordinated with the lone electron pair of the tin atom (Fig. 11).

It is very astonishing that this Lewis acid-base reaction cannot be transferred to other stable stannylenes [133, 134]. M. F. Lappert et al., as well as our group, coud not find any evidence for such adducts: neither $Sn[CH(SiMe_3)_2]_2$ (*14*) reacted with BF_3 nor $Me_2Si(NCMe_3)_2Sn$ (*1*) with Al_2Cl_6, the components of the reaction mixture being recovered unchanged. On the other hand, C. C. Hsu and R. A. Geanangel have recently reported an 1:3 adduct formed between bis(dialkylamino)tin(II) and trifluroborane (*44*) [139].

$Sn(NR_2)_2 \cdot 3\,BF_3$ $\quad$ R = Me or Et

44a, b

Multinuclear NMR experiments showed two different types of BF_3 groups, one BF_3 group being coordinated with the tin atom and two equivalent ones with the nitrogen atoms. Further studies led to the conclusion that BF_3 is initially bound to the tin atom at low molar ratios of BF_3 to $Sn(NR_2)_2$. The same authors were able to demonstrate that even base—stabilized $SnCl_2$ can be bound to BF_3. Thus, in $F_3B \cdot SnCl_2 \cdot NMe_3$ e.g. a base and a Lewis acid are simultaneously coordinated with the tin atom [140].

6.2.2 Coordination with Transition Metals

As heavier analogs of carbenes [141] stannylenes can be used as ligands in transition-metal chemistry. The stability of carbene complexes is often explained by a synergetic σ,π-effect: σ-donation from the lone electron pair of the carbon atom to the metal is compensated by a π-backdonation from filled orbitals of the metal to the empty p-orbital of the carbon atom. This concept cannot be transferred to stannylene complexes. Stannylenes are poor p-π-acceptors: no base-stabilized stannylene ($SnX_2 \cdot B$, B = electron donor) has ever been found to lose its base when coordinated with a transition metal ($M \leftarrow SnX_2 \cdot B$). Up to now, stannylene complexes of transition metals were only synthesized starting from stable monomolecular stannylenes. Divalent tin compounds are nevertheless efficient σ-donors as may be deduced from the displacement reactions (17)–(20) which open convenient routes to stannylene complexes.

(17)

M = Cr, Mo, (W) $\qquad$ *45a, b, c, d*

SnX_2 = $[(Me_3Si)_2N]_2Sn$ [142], $(C_5H_5)_2Sn$ [143], $(CO)_3Fe(CONMe_2)_2Sn$ [144], $Me_2Si(NCMe_3)_2Sn$ [145].

$$M(CO)_6 + SnX_2 \xrightarrow{h\nu} (CO)_5M\text{—}SnX_2 + CO \qquad (18)$$

M = Cr, Mo $\qquad$ *45e, d*

SnX_2 = $[(Me_3Si)_2CH]_2Sn$ [133], $\quad Me_2Si(NCMe_3)_2Sn$ [145]

$$Rh(PPh_3)_3Cl + Sn[CH(SiMe_3)_2]_2 \rightarrow Rh(PPh_3)_2ClSn[CH(SiMe_3)_2]_2 + PPh_3$$

46 $\qquad$ (19) [133]

$$Rh(PPh_3)_2(C_2H_4)Cl + Sn[CH(SiMe_3)_2]_2 \rightarrow Rh(PPh_3)_2ClSn[CH(SiMe_3)_2]_2 + C_2H_4$$

47 $\qquad$ (20) [133]

In addition to these monostannylene complexes, distannylene-metal complexes have also been prepared, either in competition with monosubstituted species (Eq. (21)) or by direct synthesis (Eq. (22)).

$$M(CO)_6 + 2\,SnX_2 \xrightarrow{h\nu} (CO)_4M(SnX_2)_2 + 2\,CO \qquad (21)$$

$M = (Cr), Mo$ *48a, b*

$SnX_2 = [(Me_3Si)_2CH]_2Sn$ [133], $MeSi(NCMe_3)_2Sn$ [145]

$$(CO)_4M\ (\text{norbornadiene}) + 2\,SnX_2 \rightarrow (CO)_4M\ (SnX_2)_2 + \text{norbornadiene} \qquad (22)$$

$M = Cr, Mo$ *48a*

$SnX_2 = [(Me_3Si)_2CH]_2Sn$ [133]

While the disubstituted compounds with the diorganylstannane *48a* exhibit exclusively trans structure, the cyclic diazastannane prefererably displays cis conformation as deduced from the IR-spectra of *48b*.

The structures of these stannylene complexes closely resemble those of carbene complexes. In Fig. 12 the crystal structure of the stannylene complex *4* is displayed; the tin atom, the two carbon and the chromium atoms are equiplanar [30].

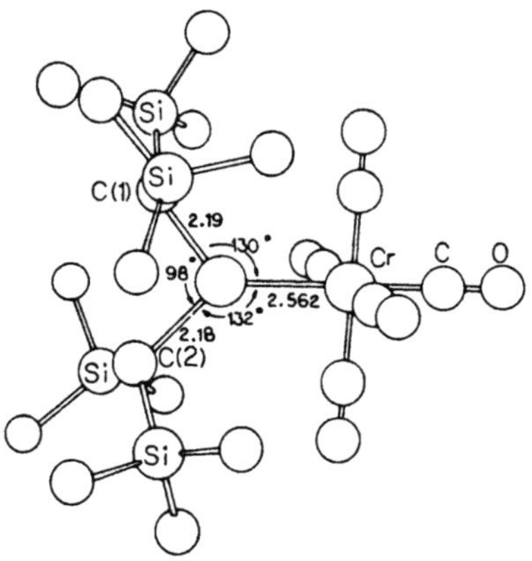

Fig. 12. Crystal structure of $[(Me_3Si)_2CH]_2Sn{-}Cr(CO)_5$ (*45*). Reprinted with permission from J. Chem. Soc., Chem. Commun. *1974*, 893. Copyright by The Chemical Society

Base-stabilized stannylenes have been known to form complexes with transition metals before stable stannylenes were detected. They are synthesized by a reaction similar to process (17) or by reduction of Sn(IV) compounds according to Eq (23) [146]:

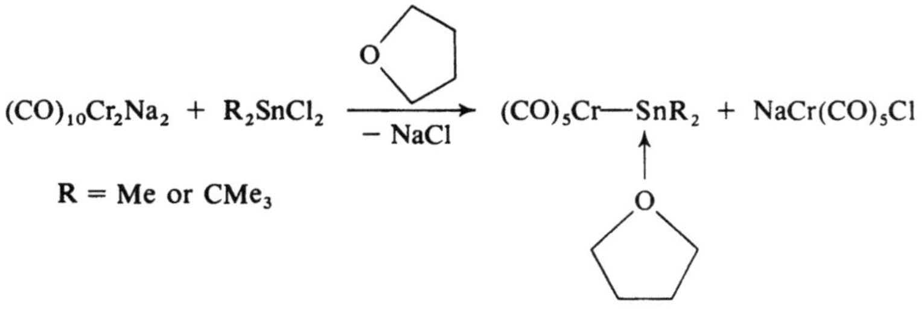

(23)

49

Often, even equilibria between μ-SnR_2-bound species and base-stabilized adducts occur: [147)]

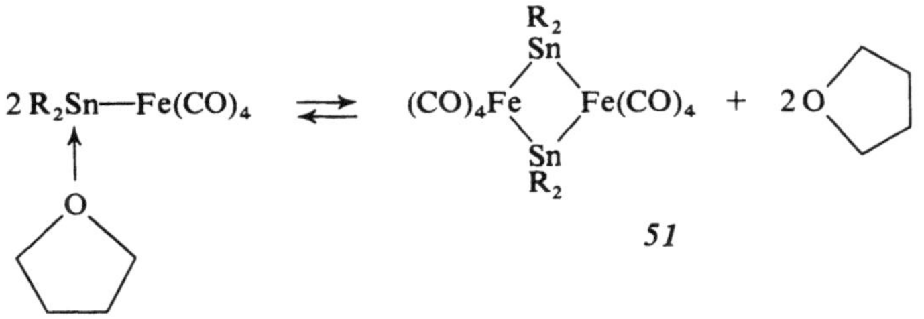

(24)

51

50

Some further selected examples of base-stabilized stannylene-metal complexes are listed in the following:

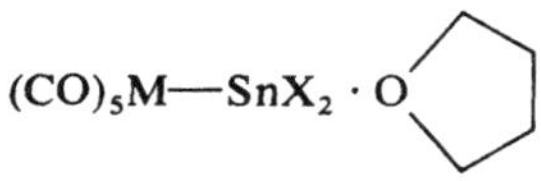

M = Cr, Mo, W
X = Cl, Br, I

52

$[(CO)_5M{-}Sn(NMe)_2]_2$

M = Cr, Mo, W

54

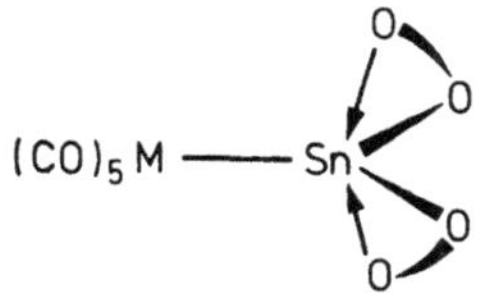

M = Cr, Mo, W

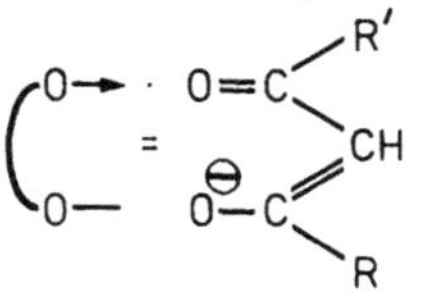

53

R = R′ = Me or CF_3;
R = Me, R′ = Ph or CF_3

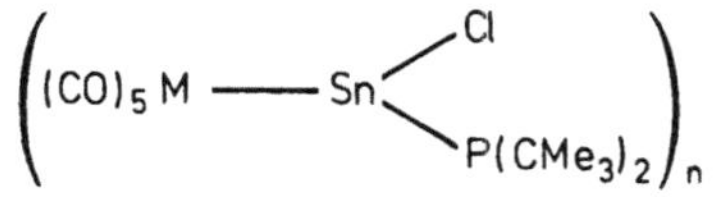

M = Cr, Mo, W

55

While *52* stands for simple base-stabilized complexes [148)], *53* is an example of a chelated base-stabilized stannylene coordinated with a transition metal [149)]. *54* [144)] and *55* [150)] illustrate that the aggregation state of the stannylene remains unchanged in the complexes.

An X-ray structural analysis of $(CO)_5CrSn(CMe_3)_2(NC_5H_5)$ confirms [33)] that the tin atom (tetrahedrally distorted) forms four bonds with neighboring atoms, the Sn—Cr bond length (265.4(3) pm) being larger than in the base-free complex *4*

(256.2 pm [30]). Besides the examples cited so far, the charged SnX_3^--ligand has attracted considerable attention in transition metal chemistry. (For more details see Ref. [151].)

6.3 Insertion Reactions of Stannylenes (Oxidative Addition)

Bivalent tin compounds can easily be transformed to fourvalent tin compounds by oxidizing agents. Mechanistically, this reaction can be understood as an insertion of a six-electron system into a two-electron bond, resulting in a tetrahedrally tetra-coordinated tin atom. This process is often also regarded as an oxidative addition, a distinction being made between additions to σ-bonds and to π-bonds.

6.3.1 Addition to Single Bonds

Stable stannylenes, base-stabilized stannylenes and unstable stannyles are known to react easily with two-electron bonds, e.g. with molecule Y—Z:

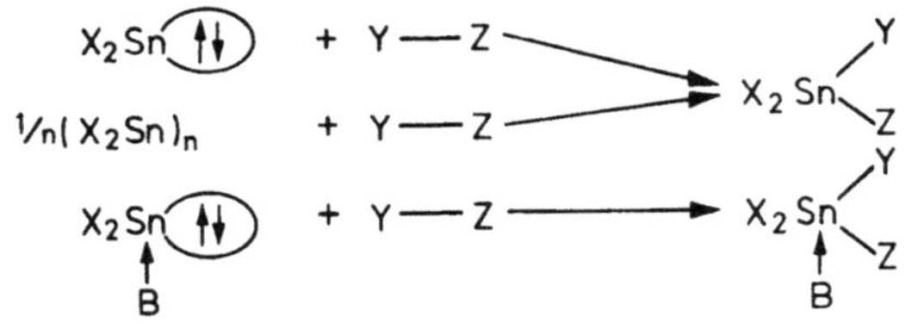

(25)

Atoms Y and Z may be identical as in halogens [133, 152, 153], in disulfanes [154], elemental sulfur [155], and polynuclear transition metal compounds [143, 156–158]. They can be different as in organic halides [86, 133, 159–163], hydrogen halides [133], transition metal halides [133, 157], transition metal hydrides [133, 164, 165], transition metal alkyles [133, 157] or in Grignard reagents [166]. Some representative compounds, which have been obtained by reactions (25) are listed below, illustrate that the substituent X bonded to the original stannylene can be varied within a great range:

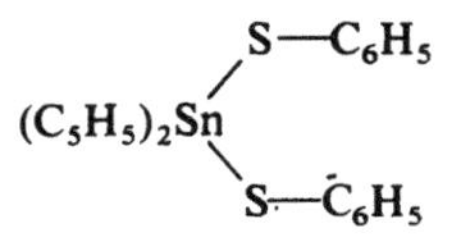

56

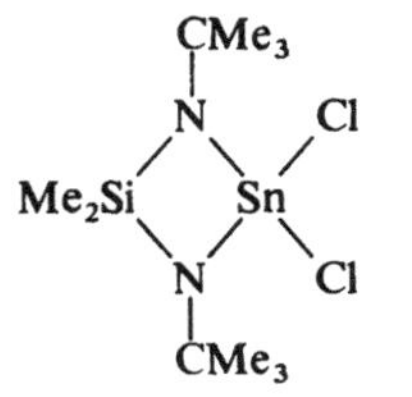

57

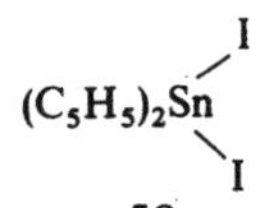

58

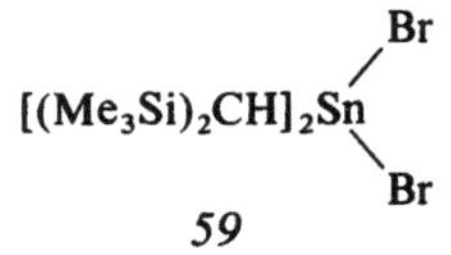

59

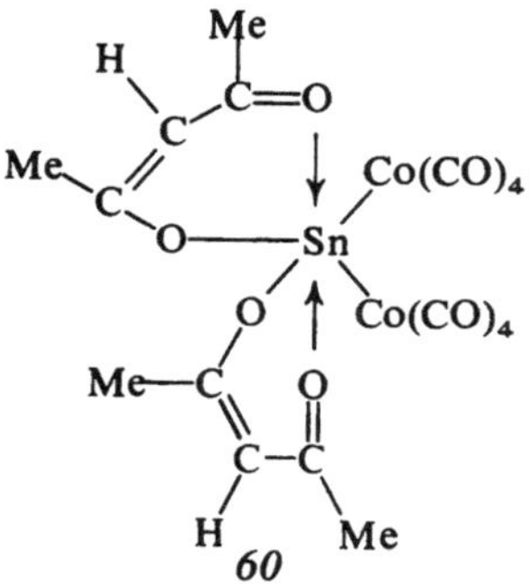

60

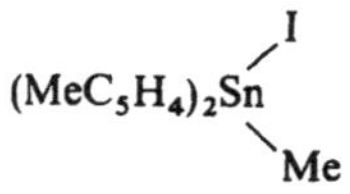

61

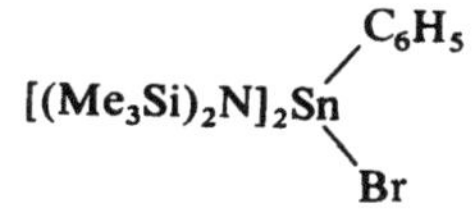

62

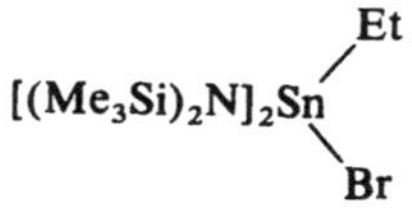

63

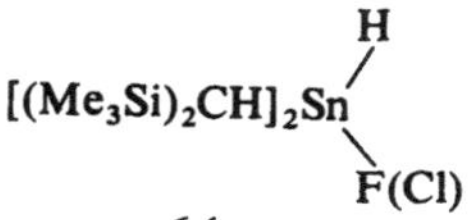

64

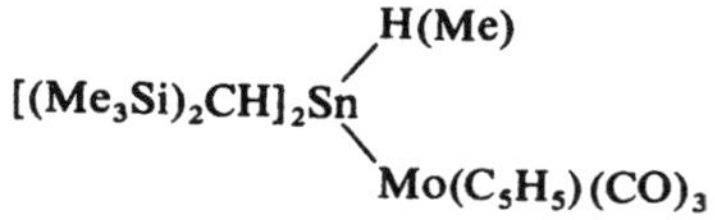

65

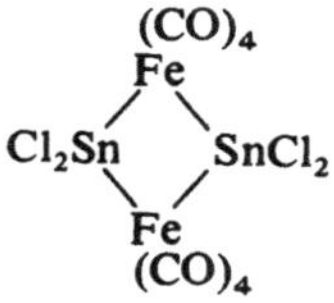

66

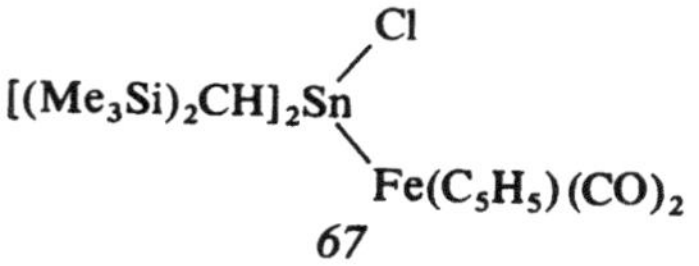

67

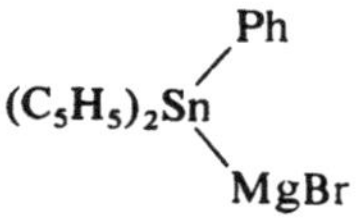

68

The reaction of the cyclic diazastannylene *1* with sulfur (S_8) in benzene yields the dispiro compound (*69*) [155]:

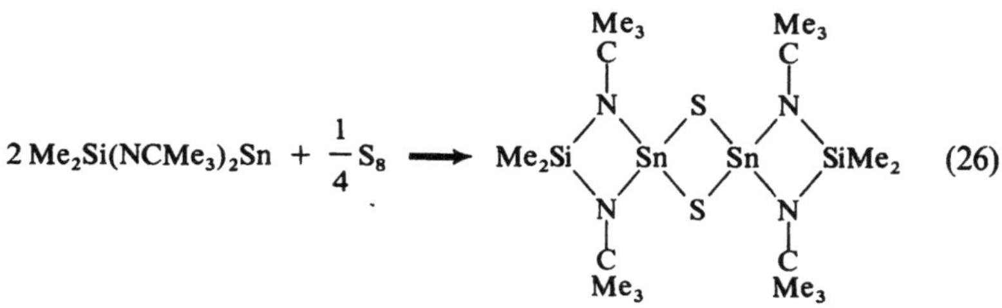

69

Another very interesting reaction involving insertion of dicyclopentadienylstannylene into metal-hydrogen bond with displacement of its ligands has been described by J. G. Noltes et al. [165]. The resulting product was identified by X-ray structural analysis.

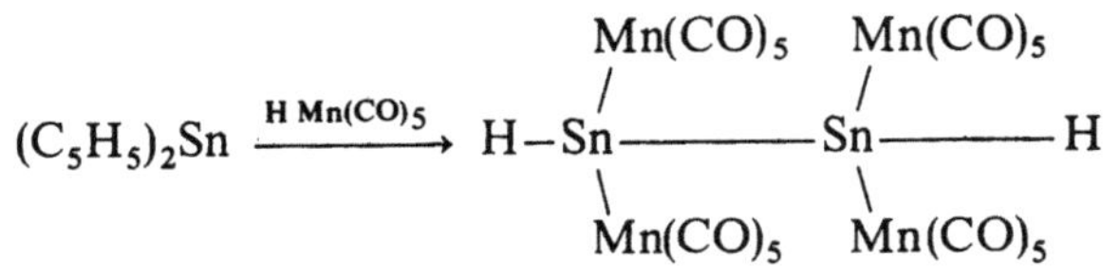

(27)

70

Insertion reactions of stannylenes, even of unstable ones, into metal-metal bonds have attracted considerable attention [156-158]. In this context, it is very astonishing that the reaction (28) between the alkyl-substituted stannylene *14* and $Fe_2(CO)_9$ does not lead to a product of type *66* ($(X_2Sn\,Fe(CO)_4)_2$ (an X-ray structural analysis indicates an Sn_2Fe_2-ring [167])) but to a three-membered ring, as determined by elemental analysis and from IR-spectral data [133].

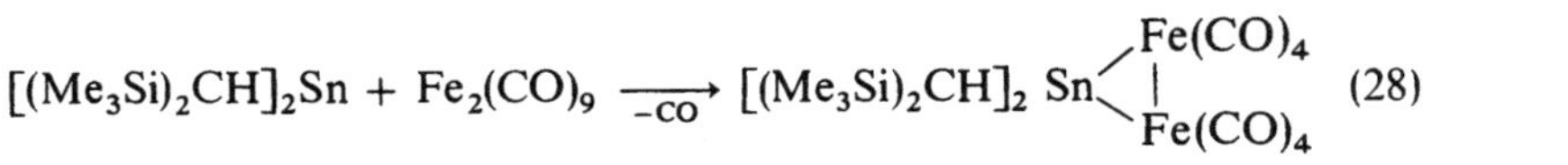

(28)

71

This reaction can also be regarded as a substitution of μ-bound carbonyl by stannylene.

Many of these reactions are of great synthetical importance as they all provide facile routes to functionally substituted tin(IV) compounds. One procedure, which is of great industrial interest, is the intermediate addition of HCl to $SnCl_2$ forming $HSnCl_3$ which reacts with C=C bonds. This type of reaction is exemplified by Eq. (29) [168]:

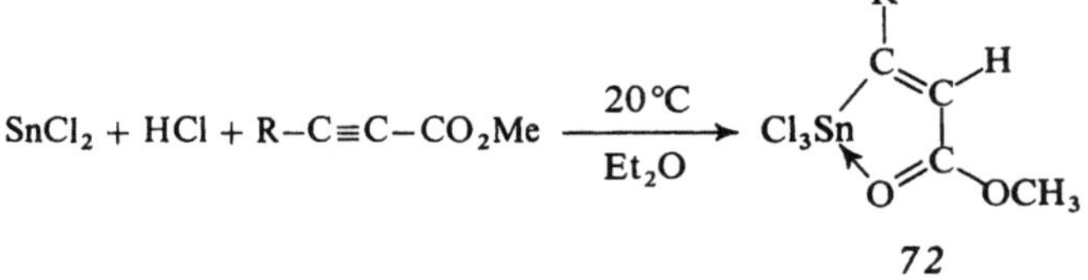

(29)

72

Nevertheless, not every σ-bond reacts with any stannylene. In principle, the Y—Z bond in Eq. (25) should be rather polar or, if it is not, the elements involved in bonding should be rather heavy (high polarizability). Up to now, no reactions involving a σ—C—C- or —N—N-bond have been reported.

Mechanistical studies on the reaction of stable stannylenes with organic halides have been performed by M. F. Lappert and his group [161-163]. On the basis of ESR spectroscopic data they proposed a radicalic pathway for this reaction. Initially, one electron of the stannylene is transferred to the organic halide the halogen of which is then added to the tin atom according to Eq. (30) (Y = halogen, X = bulky organic group).

$$SnX_2 + R'Y \rightarrow \dot{S}nX_2^{\oplus} + R'Y^{\ominus} \rightarrow \cdot SnX_2Y + \cdot R' \qquad (30)$$

6.3.2 Addition to Double Bonds

In contrast to the numerous reactions involving single bonds, interactions of stannylenes with double bonds have not extensively been studied. There are only two cases known where addition of a monomolecular stable stannylene to a double bond system takes place (Eqs. (31) [154] and (32) [133]; see also Ref. [169])

$$X_2Sn + EtO_2C{-}C{\equiv}C{-}CO_2Et \rightarrow [EtO_2C{-}\overset{|}{C}{=}\underset{\substack{|\\ SnX_2\\ |}}{C}{-}CO_2Et]_n \qquad (31)$$

$n = 2$ or 3. $X = C_5H_5$

73

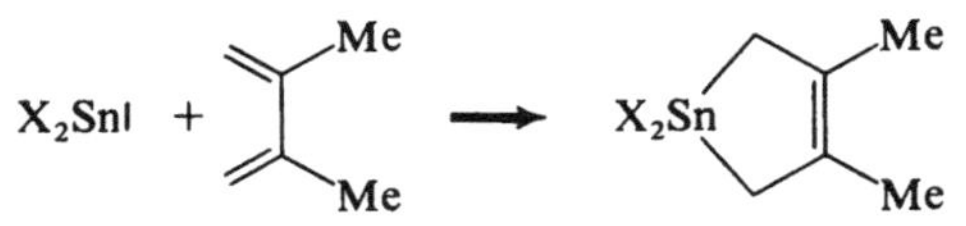

(32)

$X = [CH(SiMe_3)_2]$ *74*

It should be noted that dicyclopentadienyltin does not give any detectable reaction with 2,3-dimehtyl-1,3-butadiene, in contrast to the dialkyltin compound [154]. According to Eq. (31) 6- or 9-membered ring compounds are formed. The dicyclopentadienyl compound in Eq. (31) can also be replaced by the unstable stannylens [154].

More studies have been concentrated on the reaction of stannylenes with molecular oxygen (which is of course a limiting-case with respect to the classification as a double bond). Eq. (33) reflects the data of a variety of experiments [86, 87, 93, 133, 153].

$$X_2Sn + \frac{1}{2}O_2 \rightarrow \frac{1}{n}(X_2SnO)_n \qquad (33)$$

75a, b, c, d

$X = [CH(SiMe_3)_2], [N(SiMe_3)_2], C_5H_5, Cl$

For any ligand X of Eq. (33) a polymer *75* of definite composition but of completely unknown structure is formed. The cyclic diazastannylene *1* reacts differently. Besides the tetraazastannane *76* a crystalline solid is formed the structure of which has been

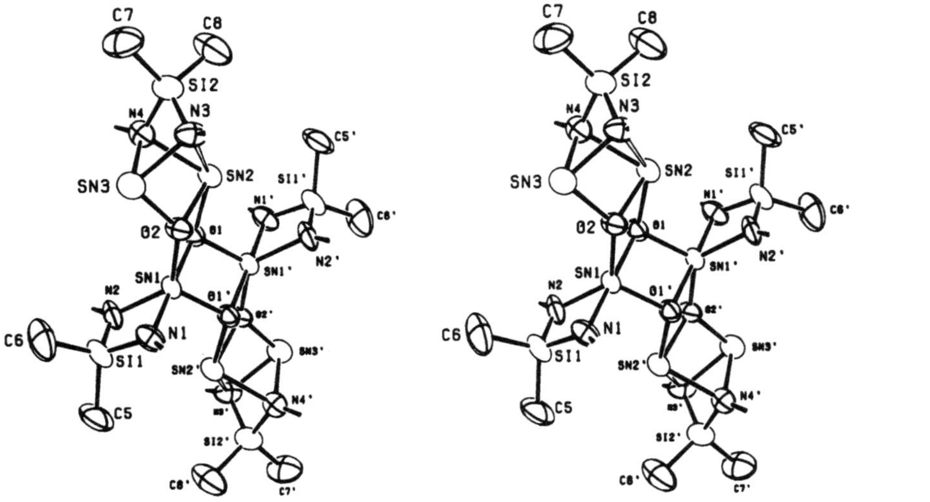

Fig. 13. A stereographic view of the crystal structure of $[Me_2Si(NCMe_3)_2Sn_2O]_2 \cdot [Me_2Si(NCMe_3)_2SnO]_2$ (*77*). Reprinted with permission from Z. Anorg. Allg. Chem. *459*, 211 (1979). Copyright by J. A. Barth Verlag

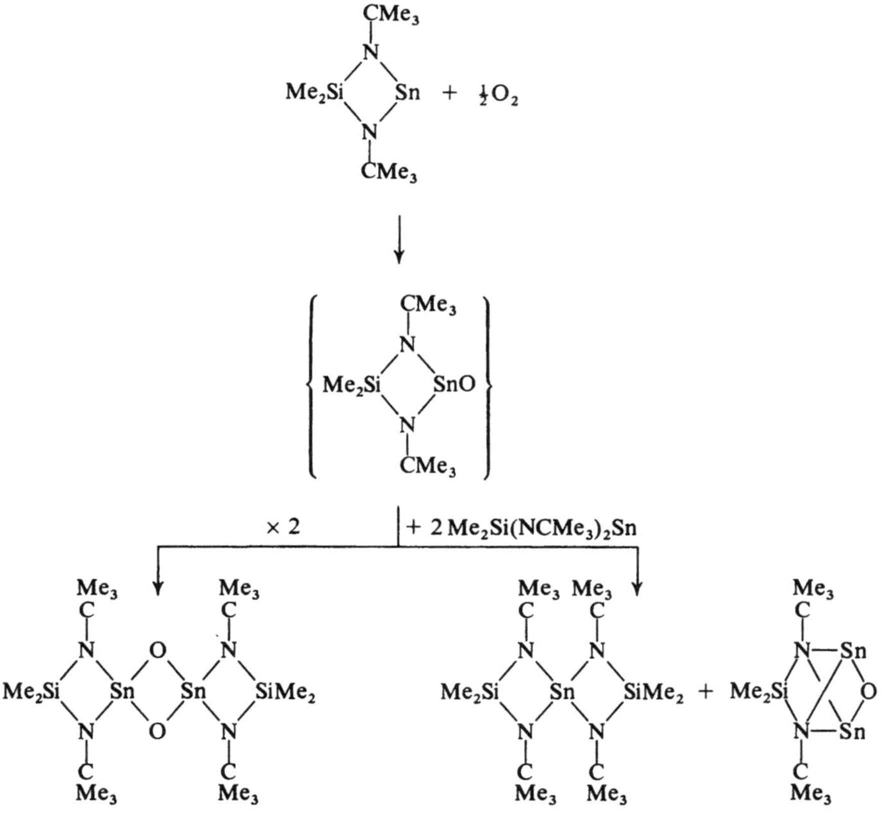

(34)

A　　　*76*　　　B

determined by X-ray structural analysis (Fig. 13). The reaction route is described by the scheme above (Eq. (34)) [170].

The dispiro compound A reacts with 2 cage molecules B to form the complex molecule *77* displayed in Fig. 13. The intermediate in brackets cannot be isolated. In contrast to the reaction of the same stannylene with sulfur (Eq. (26)) the dispiro compound A cannot be isolated seperately. The mechanism of reaction (34) may of course be more complicated. The cage molecule B is discussed in more detail in Section 6.5. It should be noted that in 77 six tin atoms of two different oxidation states are combined.

6.4 Reactions of Stannylenes with Participation of Ligands

6.4.1 SnX_2 Displaying Dihapto-Ligand Properties

In Chapter 4 a variety of stannylenes have been assembled which are characterized by a coordination of type A or B.

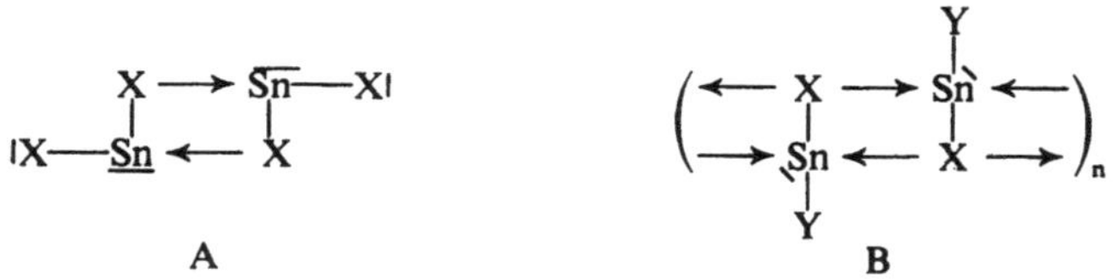

A B

The dimeric structure, which is common to the compounds *18*, *26*, *29*, *32*, can be attributed to a double intermolecular Lewis acid-base interaction, one of the substituents at the tin atom displaying base properties. The second substituent, which exhibits the same chemical properties as the first one, is not engaged in the formation of the dimer. In the case of compound type B, $(C_5H_5)ClSn$ being an illustrative representative [107], the substituent X disposes of several electron pairs and hence acts as a bifunctional base, while Y is again a terminal group. In structures A and B the stannylene displays dihapto-ligand properties, the tin atom acting as an acid and the substituent as a base.

Besides these intermolecular adducts an example for an intramolecular adduct has recently been reported. Thus, a ten-membered ring *78*, which contains two stannylene units, has been synthesized [101] (Eq. (35)).

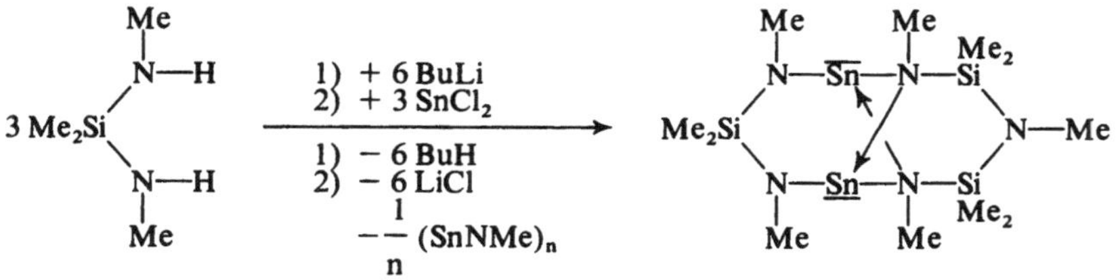

(35)

78

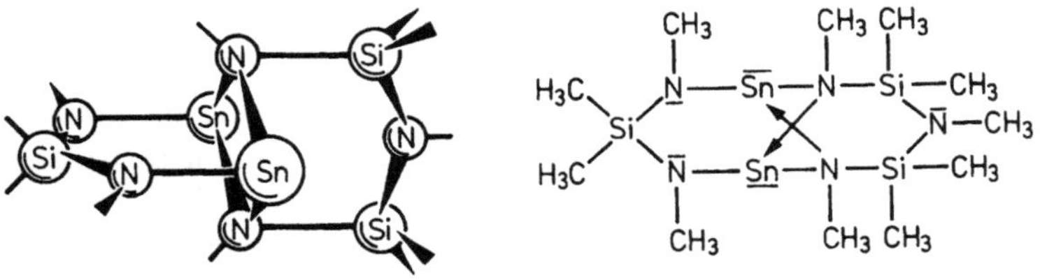

Fig. 14. Molecular structure of $(Me_2Si)_3(NMe)_5Sn_2$ (*78*) [101]

Each of these tin atoms acts as a Lewis acid, with two neighboring nitrogen atoms functioning as electron donors. The ten-membered ring folds up to generate a four-membered cycle bridged at the 1,3- and 2,4-positions. The structure, which can easily be deduced from high-resolution NMR spectra, is displayed in Fig. 14.

Up to now, no stable adducts of type A where one of the bridging atoms X is replaced by a chemically different base or the tin atom by another Lewis acid have been synthesized. Nevertheless, this type of adduct is believed to play an important part in ligand exchange or substitution reactions.

6.4.2 Ligand-Exchange Reactions

The synthesis of asymmetrically substituted stannylenes is most efficiently achieved by ligand exchange reactions between two stannylenes, SnX_2 and SnY_2 (see Chapter 4). For example, the stannylenes $[(Me_3Si)_2N]ClSn$ and $(C_5H_5)ClSn$ can be synthesized according to Eqs. (36) and (37) [86, 93, 94].

$$[(Me_3Si)_2N]_2Sn + SnCl_2 \rightarrow 2\,\underset{\textit{22}}{[(Me_3Si)_2N]ClSn} \qquad (36)$$

$$(C_5H_5)_2Sn + SnCl_2 \rightarrow 2\,\underset{\textit{21}}{(C_5H_5)ClSn} \qquad (37)$$

In both examples the reaction is shifted to the right side, because the unsymmetrical compound is highly associated and poorly soluble. If the diazastannylene and dicyclopentadienyltin are mixed in 1:1 ratio, an adduct is formed which is unstable and thus cannot be isolated; it decomposes to the original stannylenes (Eq. (38)) [87].

$$[(Me_3Si)_2N]_2Sn + (C_5H_5)_2Sn \rightleftarrows 2(C_5H_5)[(Me_3Si)_2N]Sn \xleftarrow[-2LiCl]{} 2(C_5H_5)ClSn + 2LiN(SiMe_3)_2 \qquad (38)$$

These findings may be explained by the general reaction sequence (39).

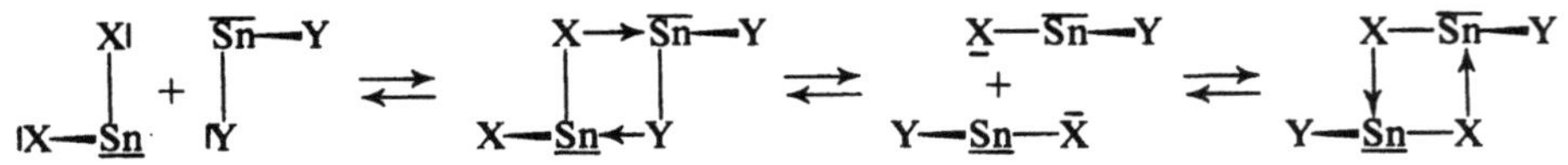

(39)

An exchange of ligands in SnX_2 and SnY_2 will occur if X is markedly more basic than Y, thus favoring the formation of an adduct formulated at the right side of reaction (39). If the basic properties of X and Y are very similar, the first equilibrium in the reaction sequence predominates and no ligand transfer is observed. Referring to our examples, reactions (36) and (37) are shifted to the right side because chlorine is a stronger donor than $[(Me_3Si)_2N]$ or cyclopentadienyl, while in Eq. (38) the two different substituents display the same basicity (the nitrogen atom in $[(Me_3Si)_2N]$ is not basic; the trimethylsilyl groups are known to reduce the basic properties considerably).

6.4.3 Ligand Substitution

As already pointed out in Chapter 4 certain stannylenes can be prepared by replacing cyclopentadienyl, dimethylamido or dimethoxy ligands by more acidic groups. In all cases, the entering ligand displaces a hydrogen atom which is transferred to the original substituent. It can be assumed (cf. also Sect. 6.5) that some of these reactions proceed via an unstable adduct in which the stannylene again acts as a dihapto ligand (Eq. (40)).

$$\begin{matrix} X \\ | \\ X{-}Sn \end{matrix} + \begin{matrix} H \\ | \\ Y \end{matrix} \longrightarrow \left\{ \begin{matrix} X \rightarrow H \\ | \quad\ \ | \\ X{-}Sn \leftarrow Y \end{matrix} \right\} \longrightarrow \begin{matrix} X{-}H \\ + \\ X{-}Sn{-}Y \end{matrix} \xrightarrow[-2\,HX]{+HY} SnY_2 \qquad (40)$$

Zuckerman et al. have extensively utilized this method in heterocyclic tin(II) chemistry [100]. In some cases, this synthesis may also be performed with tin(II) chloride, the starting hydrochloride being coordinated by the addition of an amine [120]. Free [171] and metal-bound tin(II) chloride [172] have been treated analogously with trimethyltin hydroxide to yield amorphous powders of the composition $Sn(OH)_2$ and $(CO)_5MSn(OH)_2$ (M = Cr, W) and $ClSnMe_3$. Unfortunately, no direct information on the structure of these compounds is available.

Jutzi et al. have studied reactions of bis(pentamethylcyclopentadienyl)tin with very strong acids such as HBF_4, $HAlCl_4$ or the methyl esters of CF_3COOH and CCl_3COOH [173, 174]. In these reactions only one pentamethylcyclopentadienyl group is surprinsingly replaced and a very stable cationic species $(Me_5C_5)Sn^+$ is formed:

$$(Me_5C_5)_2Sn\uparrow\downarrow + RY \rightarrow Me_5C_5R + \underset{79a,\,b,\,c,\,d}{(Me_5C_5)Sn^{\oplus}\uparrow\downarrow \quad Y^{\ominus}} \qquad (41)$$

R = H or Me

$Y^{\ominus} = BF_4^{\ominus}, AlCl_4^{\ominus}, Cl_3CCOO^{\ominus}, F_3CCOO^{\ominus}$

This outstanding behavior of bis(pentamethylcyclopendienyl)stannylene has been explained by the energetically favorable formation of the ionic compound *79* which contains the 6-membered cluster C_5Sn [174]. The structure of the boron tetrafluoride compound is illustrated in Fig. 15: the tin atom in the cation is located at the apex of a pentagonal pyramide [173].

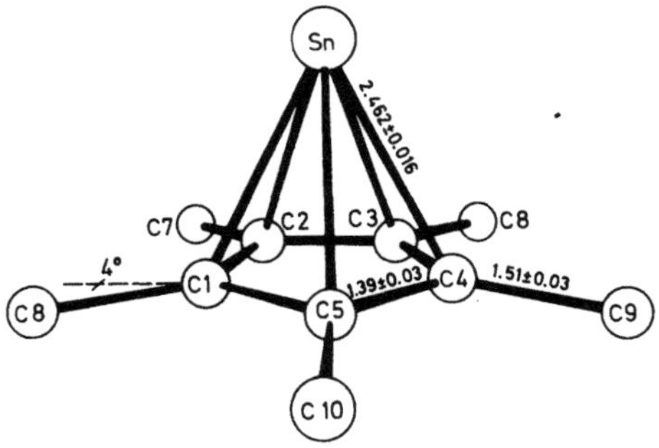

Fig. 15. Structure of the $(Me_5C_5)Sn^+$ cation in $(Me_5C_5)Sn^+BF_4^-$ (79a). Reprinted with permission from Chem. Ber. *113*, 761 (1980). Copyright by Verlag Chemie

Equation (42) combines the general features of ligand transfer reactions between tin(II) and tin(IV) compounds.

$$SnY_4 + SnX_2 \rightarrow SnY_2 + SnX_2Y_2 \tag{42}$$

The cyclic diazastannylene *1* has been found to be very suitable for this type of reaction [155] (cf. also Sect. 4.1). In Eqs. (43) and (44) the chlorine atoms of the Lewis acids are transferred to the divalent tin atom resulting in the formation of *57* and *76* and tin(II) chloride, the latter being insoluble in benzene. In (45) the solubility of the produced compounds is again important because SnS precipitates from the solution; thus, the equilibrium is shifted to the right (in Eqs. (43)–(45) R denotes tert-butyl).

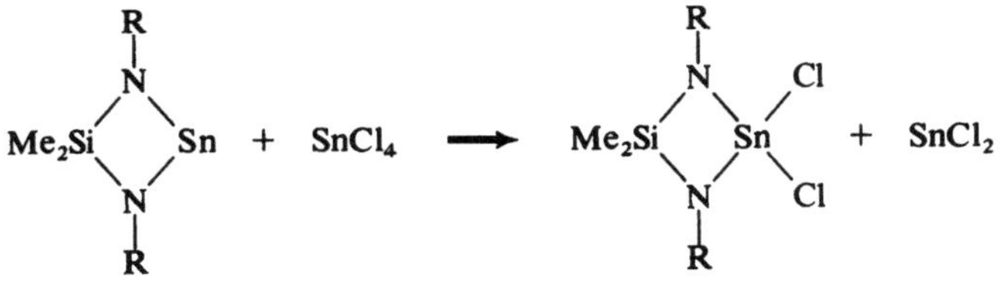

(43)

57

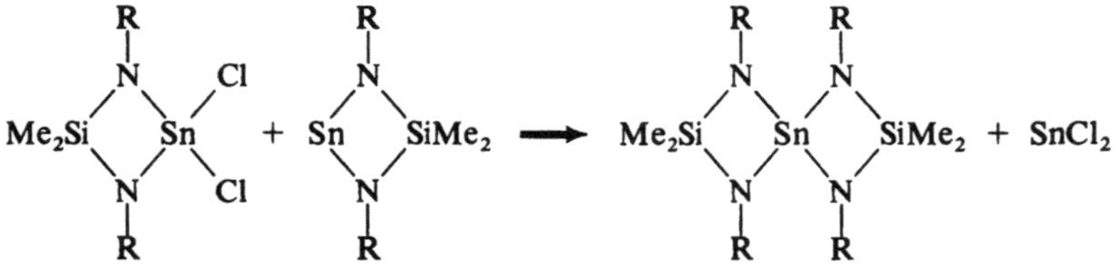

(44)

76

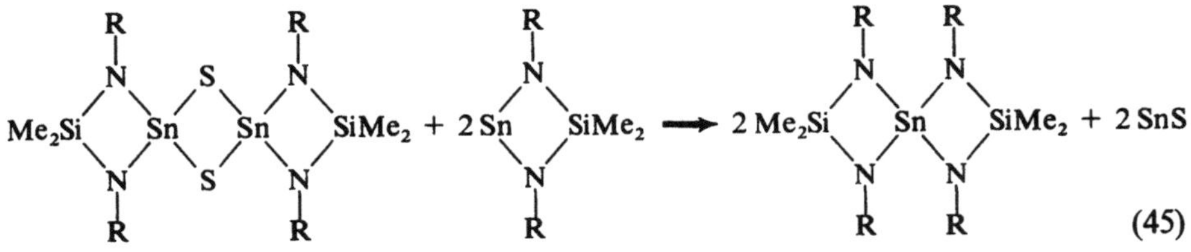

(45)

76

The rate-determinating step in reaction (45) is second order, illustrating that Lewis acid-base interactions are involved in this process. Eqs. (43)–(45) are similar to reactions (36) and (37): since the tin(II) compounds formed are highly associated (tin(II) chloride and tin(II) sulfide can be isolated as pure and large crystals), the equilibrium is shifted to the right side.

6.5 Reactions with Stannylenes SnX_2 Displaying Trihapto-Ligand Properties

From a general point of view, it seems quite unlikely that a stannylene such as SnX_2 acts simultaneously via the Lewis acidic Sn atom and the two Lewis basic ligands X as a trihapto ligand since X and Sn should first be in an appropriate position (there is free rotation around the Sn—X bonds). On the other hand, this situation must be taken into account when the ligands and the tin atom are held in a sterically fixed geometry as in cyclic derivatives of bivalent tin. Considering the cyclic diazastannylene *1* (see Sect. 4.2), the two filled p-orbitals of the nitrogen atoms as well as the empty p-orbital of the tin atom are all oriented in the same direction:

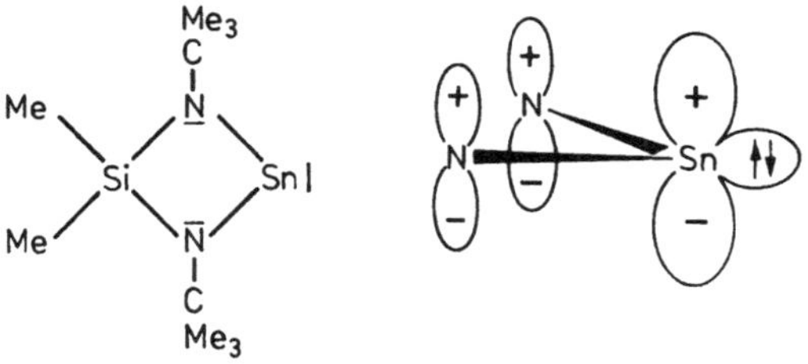

In this Section we will demonstrate that stable "adducts" of this "ligand" can be prepared and that they play an important part as intermediates. They can also be regarded as key molecules to a variety of cages.

6.5.1 Formation of Adducts

Two different "adducts" of the diazastannylene *1* acting as a trihapto ligand can be differentiated. One has the composition $Me_2Si(NCMe_3)_2Sn \cdot SnY$ and forms a stable cage molecule while the other has the general formula $Me_2Si(NCMe_3)_2Sn \cdot YH_2$ (Y in the two cases means O or N—R) and is very instable. The first adduct can be prepared according to Eqs. (46) and (47) when *1* is allowed to react with bases like H_2O or amines $H_2N—R$ [175,176].

$$2\,Me_2Si(NCMe_3)_2Sn + H_2O \xrightarrow{SnCl_2} Me_2Si(NCMe_3H)_2 + Me_2Si(NCMe_3)_2Sn_2O \cdot SnCl_2 \quad (46)$$

(*1* under the first reactant; *80* under the second product.)

$$2\,Me_2Si(NCMe_3)_2Sn + H_2N{-}R \rightarrow Me_2Si(NCMe_3H)_2 + Me_2Si(NCMe_3)_2Sn_2N{-}R \quad (47)$$

(*1* under the first reactant; *81* under the second product.)

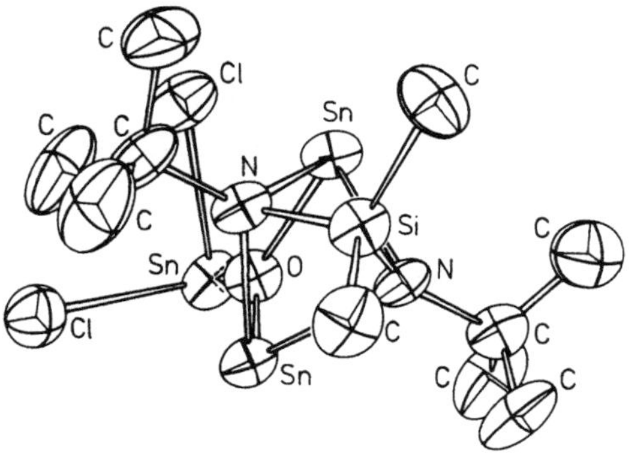

Fig. 16. The molecular structure of $Me_2Si(NCMe_3)_2Sn_2O \cdot SnCl_2$ (*80*) [175]

The second adduct is formed as an intermediate in these reactions as may be seen below. As shown by Eqs. (46) and (47) the tin atom in *1* has been replaced by two hydrogen atoms and the resulting species "SnO" resp. "SnN—R" has been trapped by a second stannylene molecule. Figure 16 describes the structure of one of the products [175]. Evidently, the cage is highly symmetrical (mm(C_{2v})symmetry), the four Sn—N bonds being equal (the $SnCl_2$ moiety of the molecule is necessary for crystallizing the compound). From a structural point of view it seems rather dubious to consider molecules of type *80* or *81* as adducts; nevertheless, it can be shown chemically. When *81* is heated in benzene solution to >200 °C, it decomposes according to 48 [176].

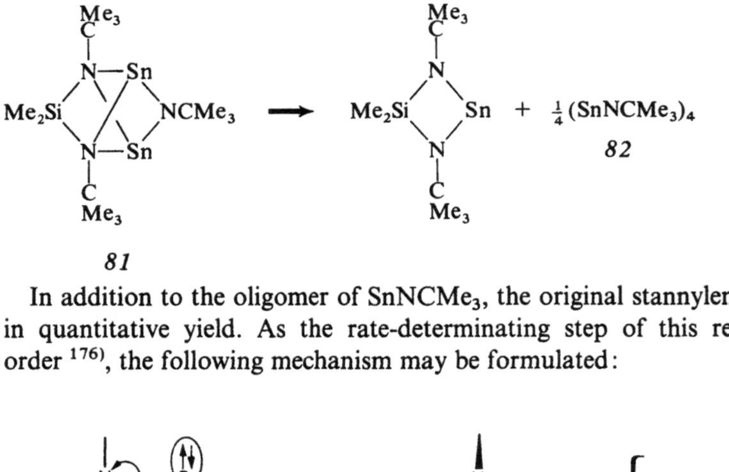

81

(49)

In addition to the oligomer of $SnNCMe_3$, the original stannylene is formed again in quantitative yield. As the rate-determinating step of this reaction is of first order [176], the following mechanism may be formulated:

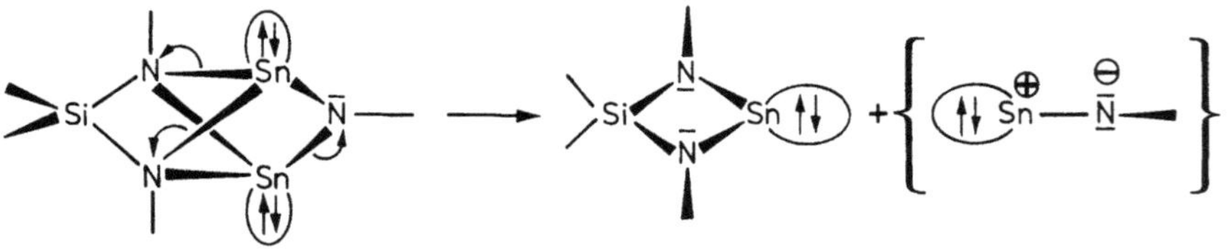

(48)

The intermediate in brackets is highly unstable and forms an oligomer as explained below. Equation (49) nicely illustrates the trihapto function of the cyclic diazastannylene opposite to the intermediate. The latter has a twofold electrophilic center at tin and a highly nucleophilic center at nitrogen.

The use of the second type of compound, in which the cyclic stannylene *1* can be considered as trihapto-coordinating, results in the same reaction 47. We have studied the mechanism of this reaction and found that the primary step is an acid-base interaction between the stannylene and the amine (Eq. (50)). It is very difficult to prove that the adduct *83* is formed in solution because of very rapid exchange of base molecules at the tin atom. We succeeded in preparing the adduct at low temperature (—70 °C) [177].

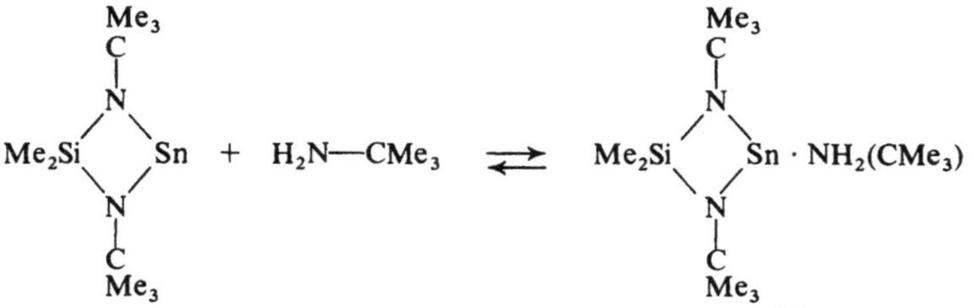

83 (50)

The structure of this adduct is displayed in Fig. 17 [177]. As should be expected (see Chapters 3 and 4) the nitrogen atom of t-butylamine is coordinated with the tin atom at nearly right angles with respect to the four-membered ring.

The rather long Sn—N distance of 246 pm as well as the nearly unchanged distances within the ring (with respect to the free stannylene) indicate that the Lewis acid-base interaction in the adduct is not very important. This parallels the findings discussed in Section 6.1.1. On the other hand, the pyramidal environment of the ring-nitrogen atoms and the narrow approach (~300 pm) of the hydrogen atoms to the latter (dashed line) is very remarkable. Again this adduct proves the trihapto properties of the stannylene: while the tin atom is coordinated with the nitrogen

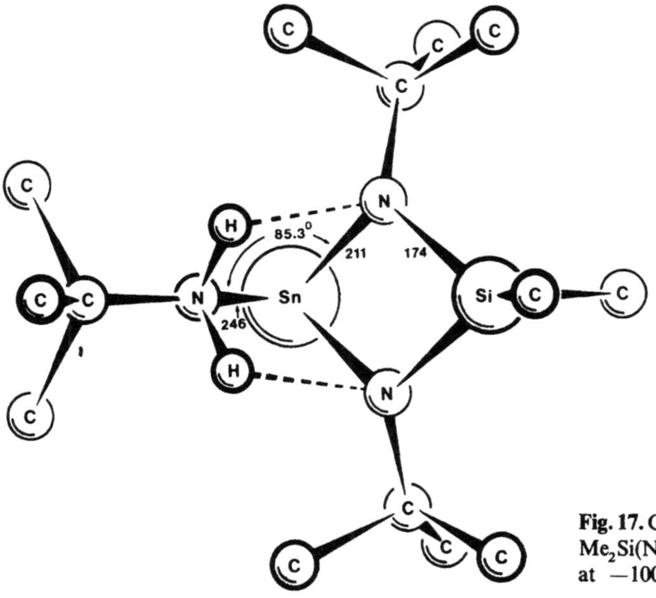

Fig. 17. Crystal structure of the adduct $Me_2Si(NCMe_3)_2Sn \cdot NH_2CMe_3$ (*83*) at —100 °C [177]

base, the electropositive hydrogen atoms slightly interact with the nitrogen atoms of the ring.

6.5.2 Synthesis of Iminostannylenes and Related Compounds

Adducts of the type $Me_2Si(NCMe_3)_2Sn \cdot H_2N{-}R$ are unstable and cleaved to $Me_2Si(N(CMe_3)H)_2$ and the intermediate $\{SnN{-}R\}$ [176,177]. It is not yet clear whether this scission is a monomolecular process, as might be inferred from the structure

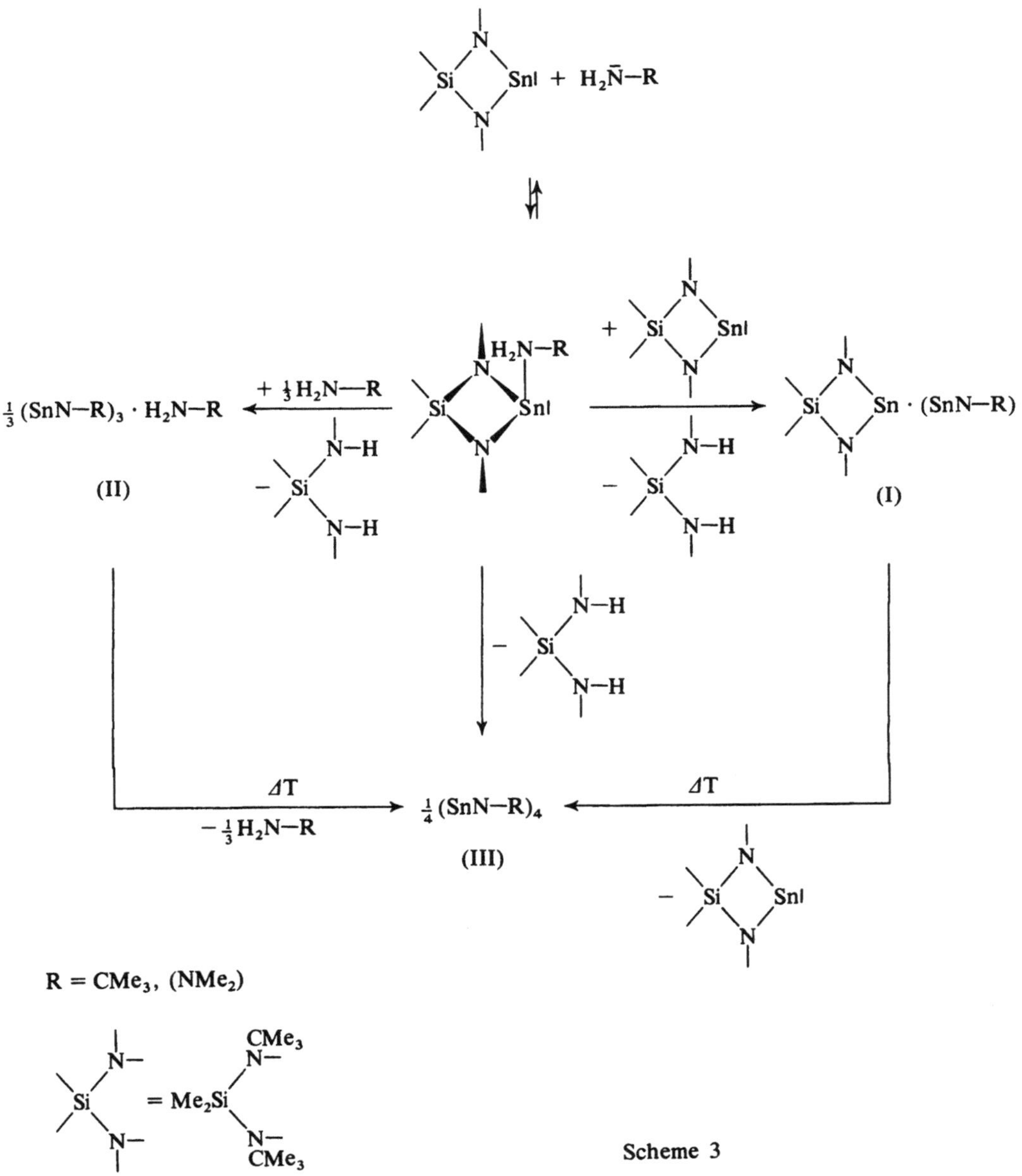

Scheme 3

(the dashed line in Fig. 17 representing the reaction coordinate of the hydrogen atoms), or a bimolecular process. Scheme 3 summarizes the different possibilities of stabilization for the fragment {SnN—R}; in all reactions a second stable fragment $Me_2Si(N(CMe_3)H)_2$ is formed.

The molecules I, II and III of Scheme 3 can be obtained, depending on the molarities of the reactants [176] or the nature of the substituent R [177]. When R is tert-butyl, thermolysis of the adduct from tert-butylamine and stannylene leads to a mixture of I and II; these compounds can be isolated in the molarities indicated in Scheme 3. In the case R = tert-butyl, compound III is not formed directly. It can however be synthesized by thermolysis of I or II at elevated temperatures. On the other hand, if R is dimethylamino, the reaction leads directly to compound III without formation of I or II [177].

The structures of I, II and III, derived from different spectra and X-ray diffraction data [32,40,175–177], are illustrated in Fig. 18. I has already been described before; II and III represent a seconorcubane-like and a cubane-like molecule, respectively. In II four nitrogen atoms form a tetrahedron, which is centered on three faces by tin atoms [40]; in III two nitrogen and tin tetrahedra interpenetrate resulting in an Sn_4N_4-cube which is highly deformed (angles at tin $\approx 80°$, at nitrogen $\approx 100°$).

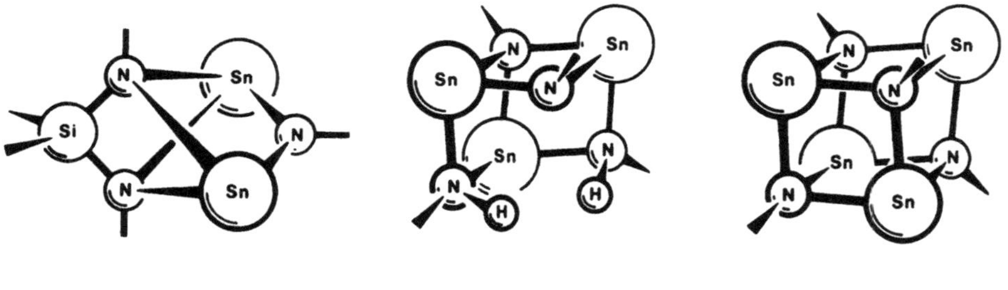

II *III*

Fig. 18. Comparison between the different structures of $Me_2Si(NCMe_3)_2Sn_2NCMe_3$ (*81*), $(Me_3CN)_4Sn_3H_2$ (*87*) and $(Me_3CN)_4Sn_4$ (*82*), resp. I, II and III of Scheme 3 [10]

The following compilation of cage compounds, which can be derived from I, II or III, illustrates the various possibilities (formal charges have been omitted!):

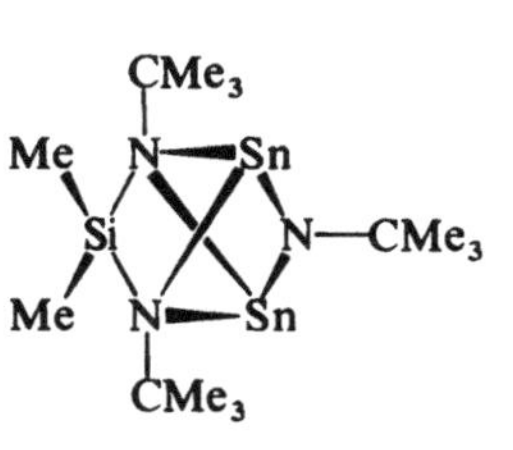

81

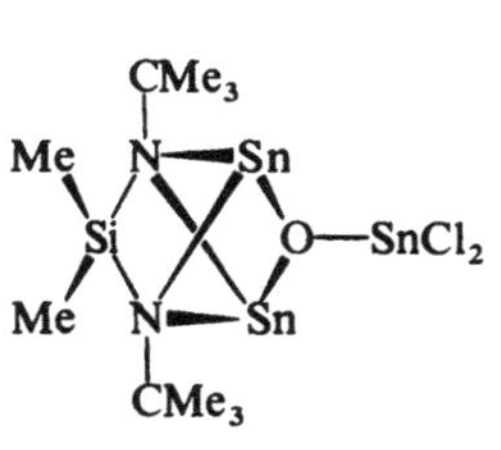

80

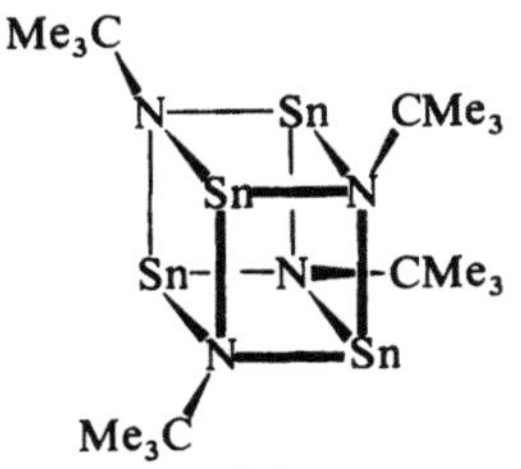

82

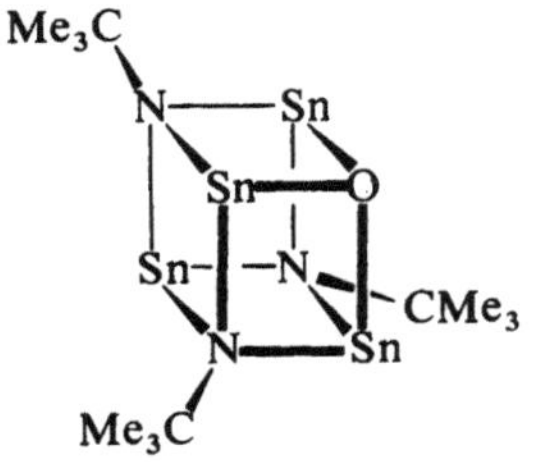

84

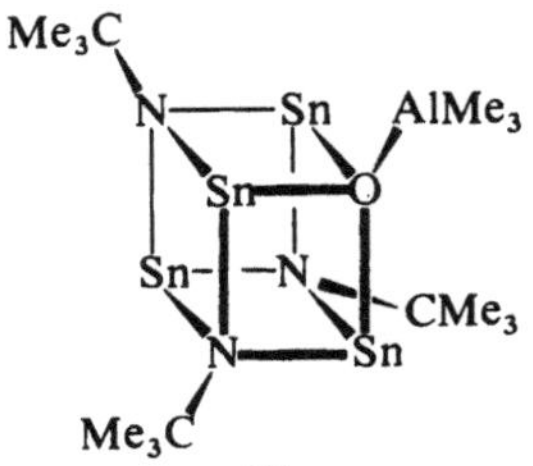

85

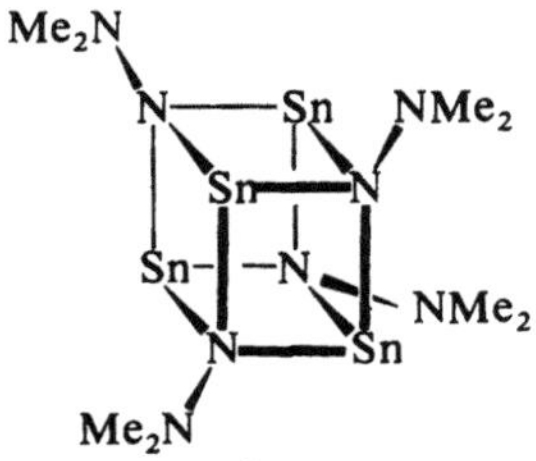

86

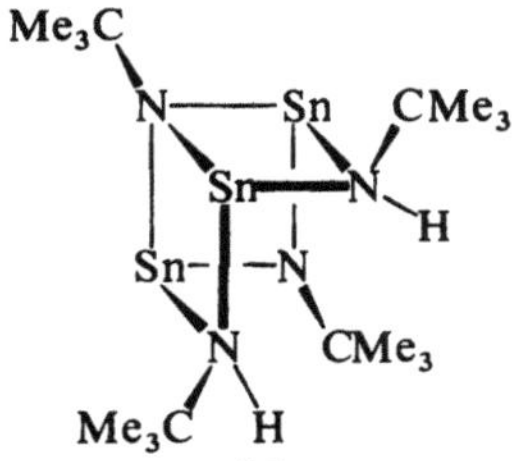

87

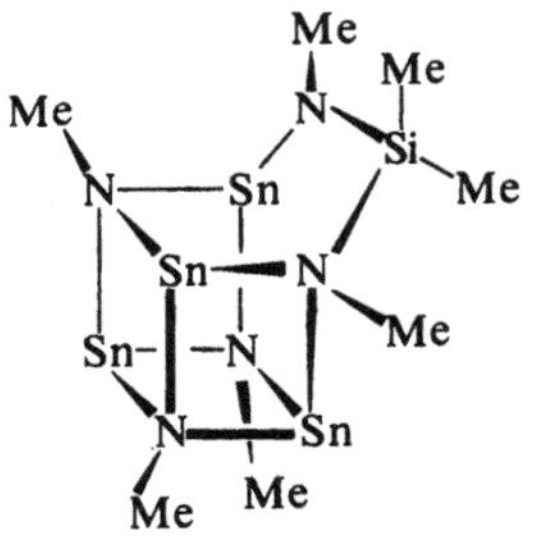

88

To illustrate the high molecular symmetry of these compounds, the molecular structure of cage *85* [178] is depicted in Fig. 19 [32]. The skeleton of *88* can be derived from an Sn_4N_4 cube enlarged at one edge by the sequence Si—N; the analog in hydrocarbon chemistry is the so-called "basketane". For further discussions of the synthesis and the structural and physical properties of these compounds the reader is referred to a review article published recently [10].

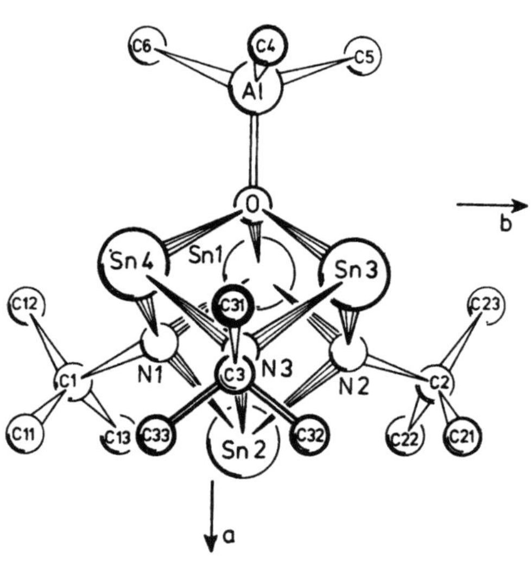

Fig. 19. Crystal structure of $(Me_3CN)_3$-$(Me_3AlO)\,Sn_4$ (*85*). Reprinted with permission from Z. Naturforsch. *36b*, 147 (1981). Copyright by Verlag Zeitschr. für Naturforsch.

We want to focus on two questions: what is the general principle for the formation of these cages and — closely related to this — how can the nature of the bonding be characterized? In contrast to isocyanides

$$R-\overset{\oplus}{N}\equiv\overset{\ominus}{C}\upharpoonleft\!\downharpoonright$$

which are stabilized by multiple bonds between the nitrogen and carbon atom, the corresponding tin compounds do not contain π-bonds (in accordance with the fact that tin is a very poor π-acceptor). As demonstrated by reaction (49), RNSn can be understood as an intermediate which may exist in several mesomeric forms:

$$\left\{\underset{A}{R-\overset{\ominus}{\underline{\overline{N}}}-\overset{\oplus}{Sn}\upharpoonleft\!\downharpoonright} \leftrightarrow \underset{B}{R\!\!\smallsetminus\!\underline{N}=Sn\upharpoonleft\!\downharpoonright} \leftrightarrow \underset{C}{R-\overset{\oplus}{N}\equiv\overset{\ominus}{Sn}\upharpoonleft\!\downharpoonright}\right\}$$

There is however no evidence for the existence of monomeric RNSn; it is either rapidly converted to a tetramer or coordinated with other functional groups (Scheme 3).

The bonding in the tetramer can be explained by two different approaches leading to the same result:

1) Taking the mesomeric form A we can combine 4 of these units in a way very similar to tetrameric thallium methoxide (omitting the formal charges) [32].

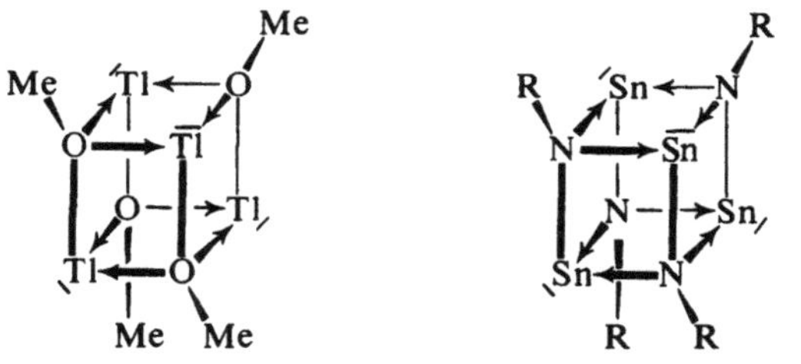

Each arrow represents a lone electron pair. These two compounds are isoelectronic with respect to their outer electron number;

2) Alternatively, in a similar way, two identical formally uncharged, four-membered rings result by combination of only two RNSn units (diazadistannylenes).

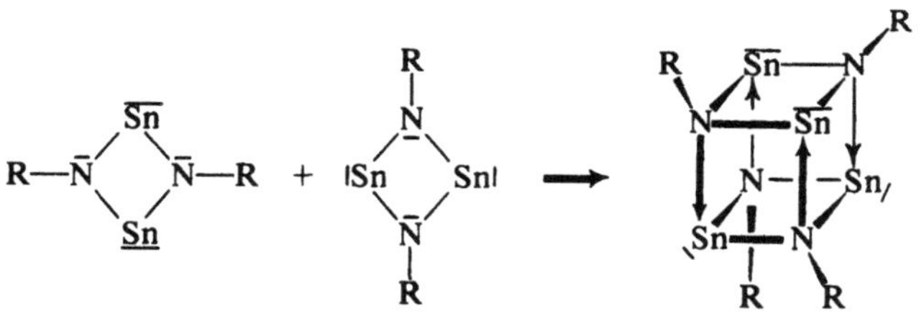

If these rings are superimposed in the correct alignment, a cube is formed due to four Lewis acid-base interactions.

The result of these approaches, which may be regarded as merely formal, is a very stable molecule, the tin atom disposing of eight electrons in the outer shell (rare-gas configuration). In contrast to the carbon atom in isocyanides, the tin atom only uses σ-bonds in these iminostannylenes (better: iminostannylines). The 12 Sn—N bonds within the cage are identical, as revealed by X-ray structural analysis [32,177]: the Sn—N bonds thus equilibrate and no difference can be found between either a homeopolar single bond or a two electron donor bond. Following the second approach, we should expect for the Sn—N bond in the cubane-like cage 66% of a single N—Sn(II) bond and 33% of a dative bond character. Taking into account independent structures of tin(II) compounds, which are in accord with these characteristics (Chapter 3 and 4), we calculate a value of 220 pm as compared with 222 pm found. The bond angles at tin with 80.5° are also in good agreement with a threefold coordinated, pyramidal tin atom (Table 7).

To sum up, the geometry of the cages can well be unterstood on the basis of structural tin(II) chemistry and the nature of the bonding can be inferred from very simple approaches. The syntheses of the cages again demonstrate the various reaction possibilities of stannylenes.

6.6 Interaction of Stannylenes with Light

Stable stannylenes are usually colored, absorbing at the beginning of the visible region of the UV spectrum (λ ≈ 400–490 nm, ε ≈ 500–1000) [108,109]. When cooled down, the color of the compound often disappears. In the case of the cyclic diazastannylene *1* this effect can be attributed to a change of the molecular structure: the monomeric molecule becomes dimeric, changing from coordination number 2 to 3 at the tin atom by intermolecular aggregation when the melt passes to the solid (Sect. 4.2). J. J. Zuckerman and P. J. Corvan have studied several SnX_2 compounds and found the following dependence: colored derivatives are always monomeric while colorless derivatives are associated [114]. It should be concluded that the color of the stannylenes depends on the coordination number at tin (in addition, it has not yet been clarified which electronic excitation is responsible for this absorption). However, this conclusion should be made with caution! This may be exemplified by $(SnN_2Me_2)_4$. Since it has a cubane-type structure, as described in the preceding chapter, it exhibits a pyramidal threefold coordination around the tin atom (X-ray structure [177]), forming dark red crystals (tin-tin interactions in the crystal or electron pairs in α-position to the tin atom might be responsible for the color).

Besides these electronic phenomena, chemical reactions can be induced when stannylenes interact with light. M. F. Lappert et al. have described that certain stable stannylenes form radicals when irradiated [179,180].

$$2\,X_2Sn \xrightarrow[-(SnX)]{h\nu} X{-}\dot{Sn}\begin{smallmatrix}\diagup X\\ \diagdown X\end{smallmatrix} \qquad (51)$$

$X = N(SiMe_3)_2$ or $CH(SiMe_3)_2$

The stability of these radicals, $\cdot SnX_3$, at ambient temperatures is very astonishing: $t^1/_2 = 3$ months for the aza-substituted and $t^1/_2 = 1$ year for the carba-substituted compound have been reported [179]. This high half-life of the radicals has been attributed to the presence of bulky substituents. According to ESR spectral data the geometry of the radicals is believed to be pyramidal ($X=N(SiMe_3)_2$) or nearly planar ($X=CH(SiMe_3)_2$) [180]. As a matter of fact these radicals represent stable compounds with tin in the oxidation state +3.

A variety of other Sn(III) radicals are known, bearing mostly organic substituents; however, they all have a very short half-life and decompose rapidly [181,182]. Reaction (51) may be regarded as a disproportionation of tin(II) and tin(I) radicals should be expected to occur in the reaction mixture (Eq. (52)).

$$2Sn(II) \xrightarrow{h\nu} Sn(I) + Sn(III) \qquad (52)$$

Evidence for these species has not yet been reported. In an experiment similar to reaction (51) dicyclopentadienyltin has been irradiated, and the reaction is believed to proceed according to Eq. (53) [183].

$$(C_5H_5)_2Sn \xrightarrow{h\nu} C_5H_5\cdot + \{(C_5H_5)Sn\cdot\} \qquad (53)$$

The cyclopentadienyl radical is well established by ESR spectroscopy while the fate of the Sn(I) radical is uncertain [183]. It should nevertheless be noted that no Sn(III) radical is formed, proving again the difference between cyclopentadienylstannylenes and other molecular tin(II) compounds.

7 References

1. Neumann, W. P.: Die Organische Chemie des Zinns, F. Enke, Stuttgart, 1967
2. Nefedov, O. M., Manakov, M. N.: Angew. Chem. *78*, 1039 (1966); Int. Ed. *5*, 1021 (1966)
3. Donaldson, J. D.: Progr. Inorgan. Chem., Vol. *8*, p. 287, Interscience, J. Wiley, New York, 1967
4. Harrison, P. G.: Coord. Chem. Rev. *20*, 1 (1976)
5. Zubieta, J. A., Zuckerman, J. J.: Progr. Inorgan. Chem., Vol. *24*, p. 251, Interscience, J. Wiley, New York, 1978
6. Cusack, P. A. et al.: Int. Tin Res. Inst., Publication No. 588 (1981)
7. Neumann, W. P., in: The Organometallic and Coordination Chemistry of Ge, Sn and Pb (M. Gielen and P. G. Harrison, (ed.), p. 51 Freund, Teil Aviv, 1978
8. Connolly, J. W., Hoff, C.: Adv. Organomet. Chem. *19*, 123 (1981)
9. Nefedov, O. M., Kolesnikov, S. P., Ioffe, A. I.: J. Organomet. Chem. Library *5*, 181 (1977)
10. Veith, M.: J. Organomet. Chem. Library *12*, 319 (1981)
11. Huheey, J. E.: Inorganic Chemistry: Principles of Structure and Reactivity, p. 620–22, Harper & Row, New York, 1972
12. Drago, R. S.: J. Phys. Chem. *62*, 353 (1958)

13. Kirmse, W.: Carbene Chemistry, 2nd Edit., Academic Press, New York 1971
14. Gillespie, R. J.: Molecular Geometry, Van Nostrand Reinhold, New York, 1972
15. Orgel, L. E.: J. Chem. Soc. *1959*, 3815
16. Tricker, M. J., Donaldson, J. D.: Inorg. Chim. Acta *31*, L445 (1978)
17. Margrave, J. L., Sharp, K. G., Wilson, P. W.: Fortschr. Chem. Forsch., Top. Current Chem. Vol. *26*, p. 1, Springer, Heidelberg, Berlin, 1972
18. Timms, P. L.: Acc. Chem. Res. *6*, 118 (1973)
19. Wyckoff, R. W. G.: Crystal Structures, 2. Ed., Vol. 1, p. 90, J. Wiley, New York, 1963
20. Rundle, R. E., Olsen, D. H.: Inorg. Chem. *3*, 596 (1964)
21. Goldberg, D. E. et al.: J. Chem. Soc., Chem. Commun. *1976*, 261
22. Lister, M. W., Sutton, L. E.: Trans. Faraday Soc. *37*, 406 (1941)
23. Akshin, P. A., Spiridonov, V. P., Khodschenko, A. N.: Zh. Fiz. Khim. *32*, 1679 (1958)
24. Fujii, H., Kimura, M.: Bull. Chem. Soc. Japan *43*, 1933 (1970)
25. Livingstone, R. L., Rao, C. N. R.: J. Chem. Phys. *30*, 339 (1959)
26. Lister, M. W., Sutton, L. E.: Trans. Faraday Soc. *37*, 393 (1941)
27. Veith, M.: Z. Naturforsch. *33b*, 7 (1978)
28. Cetinkaya, B. et al.: J. Amer. Chem. Soc. *102*, 2088 (1980)
29. Jutzi, P. et al.: Chem. Ber. *113*, 757 (1980)
30. Cotton, J. D. et al.: J. Chem. Soc., Chem. Commun. *1974*, 893
31. Bergerhoff, G., Goost, I.: Acta Cryst. *B34*, 699 (1978)
32. Veith, M., Recktenwald, O.: Z. Naturforsch. *36b*, 144 (1981)
33. Brice, M. D., Cotton, F. A.: J. Amer. Chem. Soc. *95*, 4529 (1973)
34. Olson, D. H., Rundle, R. E.: Inorg. Chem. *2*, 1310 (1963)
35. Friedel, M. K. et al.: J. Chem. Soc., Chem. Commun. *1970*, 400
36. Jeitschko, W., Sleight, A. W.: Acta Cryst *B30*, 2088 (1974)
37. Ewings, P. F. R., Harrison, P. G., King, T. J.: J. Chem. Soc., Dalton Trans. *1975*, 1455
38. Lappert, M. F. et al.: J. Chem. Soc., Chem. Commun. *1979*, 369
39. Fußstetter, H., Nöth, H.: Chem. Ber. *112*, 3672 (1979)
40. Veith, M.: Z. Naturforsch. *35b*, 20 (1980)
41. Davies, C. G. et al.: J. Chem. Soc., Dalton Trans. *1975*, 2241
42. Dewan, J. C. et al.: J. Chem. Soc., Dalton Trans. *1977*, 2319
43. Jordan, T. H. et al.: Inorg. Chem. *15*, 1810 (1976)
44. Mathew, M., Schroeder, L. W., Jordan, T. H.: Acta Cryst. *B33*, 1812 (1977)
45. Berndt, A. F.: Acta Cryst. *B30*, 529 (1974)
46. Goreaud, M., Labbé, P., Raveau, B.: Acta Cryst. *B36*, 15 (1980)
47. McDonald, R. C., Eriks, K.: Inorg. Chem. *19*, 1237 (1980)
48. Clark, S. J. et al.: Acta Cryst. *B35*, 2550 (1979)
49. Uchida, T., Kozawa, K., Obara, H.: Acta Cryst. *B33*, 3227 (1977)
50. Harrison, P. G., Haylett, B. J., King, T. J.: J. Chem. Soc., Chem. Commun. *1978*, 112
51. Christie, A. D., Howie, R. A., Moser, W.: Inorg. Chim. Acta *36*, L447 (1979)
52. Harrison, P. G., Thornton, E. W.: J. Chem. Soc., Dalton Trans. *1978*, 1274
53. Ewings, P. F. R. et al.: J. Chem. Soc.. Dalton Trans. *1976*, 1602
54. Jeitschko, W., Sleight, A. W.: Acta Cryst. *B28*, 3174 (1972)
55. van Remoortere, F. P. et al.: Inorg. Chem. *10*, 1511 (1971)
56. Herak, R. et al.: J. Chem. Soc., Dalton Trans. *1978*, 566
57. Berndt, A. F., Lamberg, R.: Acta Cryst. *B27*, 1092 (1971)
58. Donaldson, J. D., Puxley, D. C.: Acta Cryst. *B28*, 864 (1972)
59. Hauge, R. H., Hastie, J. W., Margrave, J. L.: J. Mol. Spectroscop. *45*, 420 (1973)
60. McDonald, R. C., Ho-Kuen Hau, H., Eriks, K.: Inorg. Chem. *15*, 762 (976)
61. Donaldson, J. D., Laughlin, D. R., Puxley, D. C.: J. Chem. Soc., Dalton Trans. *1977*, 865
62. Vilminot, S., Granier, W., Cot, L.: Acta Cryst. *B34*, 35 (1978)
63. Bergerhoff, G., Goost, L., Schultze-Rhonhof, E.: Acta Cryst. *B24*, 803 (1968)
64. Bergerhoff, G., Goost, L.: Acta Cryst. *B26*, 19 (1970)
65. Geneys, C., Vilminot, S.: Rev. Chim. Miner. *14*, 395 (1977)
66. McDonald, R. R., Larson, A. C., Cromer, D. T.: Acta Cryst. *17*, 1104 (1964)
67. Denes, G. et al.: J. Solid State Chem. *30*, 335 (1979)
68. Iglesias, J. E., Steinfink, H.: Acta Cryst. *B29*, 1480 (1973)

69. Mootz, D., Puhl, H.: Acta Cryst. *23*, 471 (1967)
70. Potenza, J., Mastropaola, D.: Acta Cryst. *B29*, 1830 (1973)
71. Hoskins, B. F., Martin, R. L., Rohde, N. M.: Austral. J. Chem. *29*, 213 (1976)
72. Ewings, P. F. R., Harrison, P. G., King, T. J.: J. Chem. Soc., Dalton Trans. *1976*, 1399
73. Lefferts, J. L. et al.: J. Am. Chem. Soc., in press
74. Lefferts, J. L. et al.: Angew. Chem. *92*, 326 (1980); Int. Ed. Engl. *19*, 309 (1980)
75. Dittman, G., Schäfer, H.: Z. Naturforsch. *29b*, 312 (1974)
76. Jumas, J. C. et al.: Rev. Chim. Miner. *16*, 48 (1979)
77. Rodesiler, P. F., Auel, T., Amma, E. L.: J. Amer. Chem. Soc. *97*, 7405 (1975)
78. Weininger, M. S., Rodesiler, P. F., Amma, E. L.: Inorg. Chem. *18*, 751 (1979)
79. Braun, R. M., Hoppe, R.: Angew. Chem. *90*, 475 (1978); Int. Ed. Engl. *17*, 449 (1978)
80. Grugel, C., Neumann, W. P., Seifert, P.: Tetrahedron Lett. *1977*, 2205; Kozima, S., Kobayashi, K., Kawanisi, M.: Bull. Chem. Soc. Japan *49*, 2837 (1976)
81. Grugel, C., Neumann, W. P., Schriewer, M.: Angew. Chem. *91*, 577 (1979); Int. Ed. *18*, 543 (1979)
82. Fischer, E. O., Gruber, H.: Z. Naturforsch. *11*, 423 (1956)
83. Jutzi, P., Kohl, F.: J. Organomet. Chem. *164*, 141 (1979)
84. Davidson, P. J., Lappert, M. F.: J. Chem. Soc., Chem. Commun. *1973*, 317
85. Corvan, P. J., Zuckerman, J. J.: Gov. Rep. Announce. Index (U.S.) *19*, 89 (1979)
86. Harris, D. H., Lappert, M. F.: J. Chem. Soc., Chem. Commun. *1974*, 895
87. Schaeffer, Jr., C. D., Zuckerman, J. J.: J. Amer. Chem. Soc. *96*, 7160 (1974)
88. Veith, M.: Angew. Chem. *87*, 287 (1975); Int. Ed. Engl. *14*, 263 (1975)
89. Lappert, M. F. et al.: J. Chem. Soc., Chem. Commun. *1980*, 621
90. Du Mont, W. W., Kroth, H. J.: Angew. Chem. *89*, 832 (1977); Int. Ed. Engl. *16*, 792 (1977)
91. Ewings, P. F. R., Harrison, P. G.: J. Chem. Soc., Dalton Trans. *1975*, 2015
92. Du Mont, W. W.: Inorg. Chim. Acta *29*, L195 (1978)
93. Harrison, P. G., Zuckerman, J. J.: J. Amer. Chem. Soc. *91*, 6885 (1969)
94. Bos, K. D., Bulten, E. J., Noltes, J. G.: J. Organomet. Chem. *39*, C52 (1972)
95. Ewings, P. F. R., Harrison, P. G., Fenton, D. E.: J. Chem. Soc., Dalton Trans. *1975*, 821
96. Cornwell, A. B., Harrison, P. G.: J. Chem. Soc., Dalton Trans. *1975*, 1722
97. Bos, K. D. et al.: Inorg. Nucl. Chem. Letters *9*, 961 (1973)
98. Ewings, P. F. R., Harrison, P. G.: J. Chem. Soc., Dalton Trans. *1975*, 1717
99. Zeldin, M., Gsell, R.: Synth. React. Inorg. Metal-Org. Chem. *6*, 11 (1976)
100. Honnick, W. D., Zuckerman, J. J.: Inorg. Chem. *17*, 501 (1978)
101. Veith, M., Grosser, M., Recktenwald, O.: J. Organomet. Chem. *216*, 27 (1981)
102. Honnick, W. D., Zuckerman, J. J.: Inorg. Chem. *18*, 1437 (1979)
103. Petz, W., Jonas, A.: J. Organomet. Chem. *120*, 423 (1976)
104. Dave, L. D., Evans, D. F., Wilkinson, G.: J. Chem. Soc. *1959*, 3684
105. Harrison, P. G., Healy, M. A.: J. Organomet. Chem. *51*, 153 (1973)
106. Almenningen, A., Haaland, A., Motzfeldt, T.: J. Organomet. Chem. *7*, 97 (1967)
107. Bos, K. D. et al.: J. Organomet. Chem. *99*, 71 (1975)
108. Davidson, P. J., Harris, D. H., Lappert, M. F.: J. Chem. Soc., Dalton Trans. *1976*, 2268
109. Gynane, M. J. S. et al.: J. Chem. Soc., Dalton Trans. *1977*, 2004
110. Foley, P., Zeldin, M.: Inorg. Chem. *14*, 2264 (1975)
111. Du Mont, W. W., Neudert, B.: Z. anorg. allg. Chem. *436*, 270 (1977)
112. Veith, M.: Z. Naturforsch. *33b*, 1 (1978)
113. Hänssgen, D., Kuna, J., Ross, B.: Chem. Ber. *109*, 1797 (1976)
114. Corvan, P. J., Zuckerman, J. J.: Inorg. Chim. Acta *34*, L255 (1979)
115. Amberger, E., Kula, M. R.: Chem. Ber. *96*, 2562 (1963)
116. Morrison, J. S., Haendler, H. M.: J. Inorg. Nucl. Chem. *29*, 393 (1967)
117. Tzschach, A.: J. Organomet. Libr. *12*, 293 (1981); 3. Int. Symp. Org.-Met. Coord. Chem. of Ge, Sn and Pb, Dortmund 1980
118. Cotton, J. D. et al.: J. Chem. Soc., Dalton Trans. *1976*, 2286
119. Cocks, G. T., Zuckerman, J. J.: Inorg. Chem. *4*, 592 (1965)
120. Hall, W. T., Zuckerman, J. J.: Inorg. Chem. *16*, 1239 (1977)
121. Cornwell, A. B., Cornwell, C. A., Harrison, P. G.: J. Chem. Soc., Dalton Trans. *1976*, 1612
122. Du Mont, W. W., Lefferts, J. L., Zuckerman, J. J.: J. Organomet. Chem. *166*, 347 (1979)

123. Cusack, P. A., Smith, P. J., Donaldson, J. D.: Inorg. Chim. Acta, Bioinorg. Chem. *46*, L73 (1980)
124. Veith, M., Lange, H.: unpublished
125. Harris, D. H. et al.: J. Chem. Soc., Dalton Trans. *1976*, 945
126. Gmelin: Handbuch der Anorganischen Chemie, Zinn, Part C5, p. 3–62, Springer: Berlin, Heidelberg, 1977
127. Deans, R. A., Geanangel, R. A.: Inorg. Chim. Acta *43*, 159 (1980)
128. Jaura, K. L., Singh, B., Chadka, R. K.: Indian J. Chem. *12*, 1304 (1974)
129. Kauffman, J. W., Moor, D. H., Williams, R. J.: J. Inorg. Nucl. Chem. *39*, 1165 (1977)
130. Du Mont, W. W. et al.: Angew. Chem. *88*, 303 (1976); Int. Ed. *15*, 308 (1976)
131. Du Mont, W. W., Neudert, B.: Z. anorg. allg. Chem. *441*, 86 (1978)
132. Hsu, C. C., Geanangel, R. A.: Inorg. Chem. *16*, 2529 (1977)
133. Cotton, J. D., Davidson, P. J., Lappert, M. F.: J. Chem. Soc., Dalton Trans. *1976*, 2275
134. Veith, M.: unpublished
135. Du Mont, W. W.: J. Organomet. Chem. *153*, C11 (1978)
136. Harrison, P. G., Zuckerman, J. J.: J. Amer. Chem. Soc. *92*, 2577 (1970)
137. Doe, J., Borkett, S., Harrison, P. G.: J. Organomet. Chem. *52*, 343 (1973)
138. Harrison, P. G., Richards, J. A.: J. Organomet. Chem. *108*, 35 (1976)
139. Hsu, C. C., Geanangel, R. A.: Inorg. Chim. Acta *34*, 241 (1979)
140. Hsu, C. C., Geanangel, R. A.: Inorg. Chem. *19*, 110 (1980)
141. Fischer, E. O.: Adv. Organomet. Chem. *14*, 1 (1976)
142. Lappert, M. F., Power, P. P.: Adv. Chem. Ser. *157*, 70 (1976)
143. Cornwell, A. B., Harrison, P. G., Richards, J. A.: J. Organomet. Chem. *108*, 47 (1976)
144. Petz, W.: J. Organomet. Chem. *165*, 199 (1979)
145. Veith, M. et al.: J. Organomet. Chem. *216*, 377 (1981)
146. Marks, T. J.: J. Amer. Chem. Soc. *93*, 7090 (1971)
147. Marks, T. J., Newman, A. R.: J. Amer. Chem. Soc. *95*, 769 (1973)
148. Uhlig, D., Behrens, H., Lindner, E.: Z. anorg. allg. Chem. *401*, 233 (1973)
149. Cornwell, A. B., Harrison, P. G.: J. Chem. Soc., Dalton Trans. *1975*, 1486
150. Du Mont, W. W., Neudert, B.: Chem. Ber. *111*, 2267 (1978)
151. Wells, A. F.: Structural Inorganic Chemistry, 4. Ed., Clarendon Press, Oxford 1975
152. Schröer, U., Albert, H. J., Neumann, W. P.: J. Organomet. Chem. *102*, 291 (1975)
153. Messin, G., Janier-Dubry, J. L.: Inorg. Nucl. Chem. Lett. *15*, 409 (1979)
154. Bos; K. D., Bulten, E. J., Noltes, J. G.: J. Organomet. Chem. *67*, C13 (1974)
155. Veith, M., Recktenwald, O., Humpfer, E.: Z. Naturforsch. *33b*, 14 (1978)
156. Cornwell, A. B., Harrison, P. G.: J. Chem. Soc., Dalton Trans. *1976*, 1608
157. Cole, B. J., Cotton, J. D., McWilliam, D.: J. Organomet. Chem. *64*, 223 (1974)
158. Cornwell, A. B., Harrison, P. G.: J. Chem. Soc., Dalton Trans. *1975*, 2017
159. Albert, H. J., Schröer, U.: J. Organomet. Chem. *60*, C6 (1973)
160. Bos, K. D., Bulten, E. J., Noltes, J. G.: J. Organomet. Chem. *99*, 397 (1975)
161. Gynane, M. J. S. et al.: J. Chem. Soc., Chem. Commun. *1976*, 256
162. Gynane, M. J. S. et al.: J. Chem. Soc., Dalton Trans. *1977*, 2009
163. Gynane, M. J. S. et al.: J. Chem. Soc., Chem. Commun. *1978*, 192
164. Hoff, C. D., Connolly, J. W.: J. Organomet. Chem. *148*, 127 (1978)
165. Bos, K. D., Bulten, E. J., Noltes, J. G.: J. Organomet. Chem. *92*, 33 (1975)
166. Harrison, P. G., Zuckerman, J. J., Noltes, J. G.: J. Organomet. Chem. *31*, C23 (1971)
167. Harrison, P. G., King, T. J., Richards, J. A.: J. Chem. Soc., Dalton Trans. *1975*, 2097
168. von Werner, K., Blank, H., Yasufuku, K.: J. Organomet. Chem. *165*, 187 (1979)
169. Meier, P. F. et al.: Inorg. Chem. *18*, 2051 (1979)
170. Veith, M., Recktenwald, O.: Z. anorg. allg. Chem. *459*, 208 (1979)
171. Honnick, W. D., Zuckerman, J. J.: Inorg. Chem. *15*, 3034 (1976)
172. Du Mont, W. W., Neudert, B.: Angew. Chem. *92*, 561 (1980); Int. Ed. *19*, 553 (1980)
173. Jutzi, P., Kohl, P., Krüger, C.: Angew. Chem. *91*, 81 (1979); Int. Ed. *18*, 59 (1979)
174. Kohl, F. X., Jutzi, P., Chem. Ber. *114*, 488 (1981)
175. Veith, M.: Chem. Ber. *111*, 2536 (1978)
176. Veith, M., Sommer, M.-L., Jäger, D.: Chem. Ber. *112*, 2581 (1979)
177. Veith, M., Schlemmer, G.: unpublished

178. Veith, M., Lange, H.: Angew. Chem. *92*, 408 (1980); Int. Ed. *19*, 401 (1980)
179. Cotton, J. D. et al.: J. Chem. Soc., Chem. Commun. *1974*, 651
180. Hudson, A., Lappert, M. F., Lednor, P. W.: J. Chem. Soc., Dalton Trans. *1976*, 2369
181. Buschhaus, H. U., Lehnig, M., Neumann, W. P.: J. Chem. Soc., Chem. Commun. *1977*, 129
182. Lehnig, M., Dören, K.: J. Organomet. Chem. *210*, 331 (1981)
183. Davies, A. G., Tse, M. W.: J. Chem. Soc., Chem. Commun. *1978*, 353

Chirality, Static and Dynamic Stereochemistry of Organotin Compounds

Marcel Gielen

Vrije Universiteit Brussel, T.W.-AOSC, Pleinlaan 2, B-1050 Brussel, Belgium
and
Université Libre de Bruxelles, Collectif de Chimie Organique Physique

Table of Contents

The methods which have been used to determine the configurational stability of organotin compounds and those which have successfully been applied to synthesize such optically active molecules are reviewed, including the chromatographic resolution of racemic mixtures.

The optical stability of organotin compounds is then discussed and the stereoselectivity (or non-stereoselectivity) of a series of substitution reactions at the tin atom is described and used to distinguish between possible mechanisms.

1 Introduction

The first examples of optically active organosilicon compounds, methyl-1-naphthylphenylsilicon chloride, hydride and methoxide

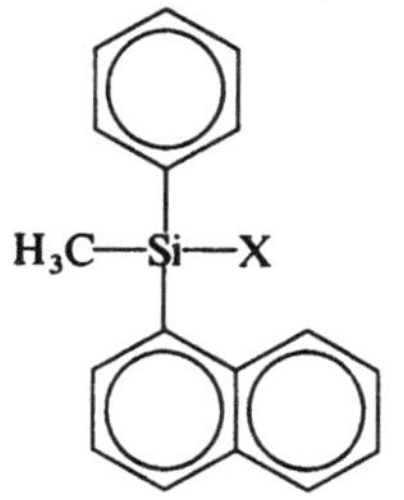

X = Cl, H, CH_3O (1)

were prepared by Sommer in 1959 [1].

The first examples of optically active organogermanium compounds [see Ref. [2]] in which the metal atom is again the only chiral center were published in 1963, by Eaborn [3] and by Brook [4]. They prepared optically pure (+)-ethyl- and (—)-methyl-1-naphthylphenylgermanium hydrides, using Sommer's resolution method [1].

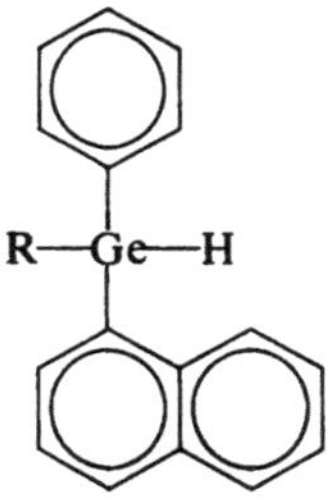

R = CH_3CH_2, CH_3 (2)

As early as 1900 Pope and Peachey [5a] reported the synthesis of (+)-methylethylpropyltin iodide, (+)- (*1*), via the corresponding (+)-camphor-β-sulfonate (*1'*)

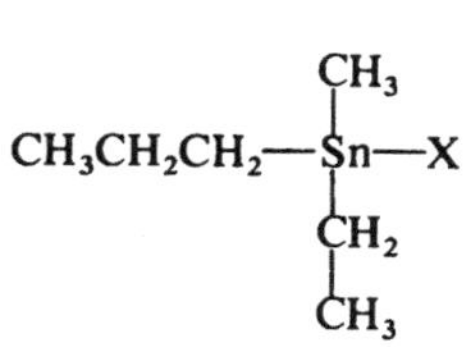

X = I; X

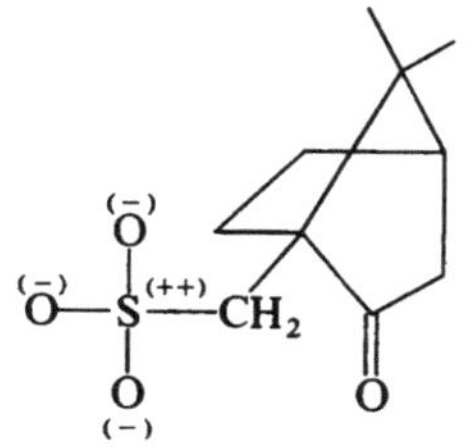

(3)

(*1*) (*1'*)

However this resolution was based on erroneous results: In 1928 Kipping [6] tried to use the same method to resolve aromatic organotin compounds and failed.

In 1935, Naumov and Manulkin [7] tried without success to isolate (—)-methylethylpropyltin iodide, (—)-(*1*). In fact, triorganotin halides are generally configurationally unstable, as shown in 1968 by Peddle and Redl [8], whereas tetraorganotin compounds appeared to be configurationally stable within the NMR time-scale (see below).

Nevetheless, Pope and Peachey's paper [5a] was still being mentioned in at least one recent review [5b] as the first example of an optically active tin compound, without Peddle and Redl's correction [8].

Extrapolating the configurational instability of triorganotin halides to other organotin compounds, Belloli [9] wrote in 1969: "*The resolution of the corresponding tin and lead enantiomers would complete the isoconfigurational series, but the finding of rapid inversion for asymmetric tin (and therefore probably also for lead) is a serious obstacle to achieve this objective*" [9].

Peddle and Redl [10] were still rather pessimistic in 1970: "*Thus while it should be possible to resolve an optically active organotin compound with four carbon-tin bonds, it seems unlikely that such a compound would be very useful in investigating the stereochemistry of substitution at the tin atom*" [10].

In 1973, Folli, Iarossi and Taddei [11] gave their opinion about Peddle and Redl's affirmation: "*As pointed out before, it should be possible to resolve optically active organotin compounds with four carbon-tin bonds, since they seem to have a high stereochemical stability. In any event, we believe that their importance in investigating the stereochemistry of substitution at tin might be more important than is generally thought, if this optical activity is obtained through a reaction which enables one to investigate the stability of the asymmetric tin atom*" [11].

2 NMR Determination of the Configurational Stability of Organotin Compounds

2.1 The Use of Diastereotopic Groups as a Probe for the Determination of the Configurational Stability at the Tin Atom

Organotin compounds with diastereotopic groups can be used to determine how rapidly tin inverts its configuration. This is shown below: if a racemic mixture M + $\bar{M}$ is examined by NMR, two distinct signals are expected for the diastereo-

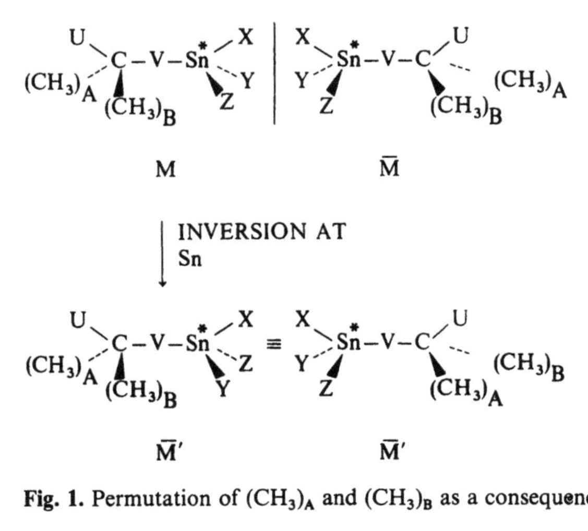

Fig. 1. Permutation of $(CH_3)_A$ and $(CH_3)_B$ as a consequence of the inversion of the tin atom

topic methyls (other diastereotopic groups may be used instead), one for $(CH_3)_A$, one for $(CH_3)_B$.

If the tin atom inverts its configuration, M is converted into $\overline{M}'$ but, because the carbon atom bearing the two diastereotopic methyl groups does not invert, the methyl A of M has been converted into the methyl B of $\overline{M}'$, and the methyl B of M, into the methyl A of $\overline{M}'$. Similarly, $(CH_3)_A$ of $\overline{M}$ is converted into $(CH_3)_B$ of M′, the mirror image of $\overline{M}'$ and $(CH_3)_B$ of M, tinto the $(CH_3)_A$ of $\overline{M}'$.

If the inversion at the tin atom occurs at a rate v such that

$$\frac{v}{[M]} = (\pi/\sqrt{2})\,\Delta v_\infty$$

where [M] is the concentration of the organotin compound, and Δv_∞, the difference in chemical shift, expressed in Herz, between $(CH_3)_A$ and $(CH_3)_B$ in the absence of any exchange, then the coalescence of the signals of A and B will be observed.

It is therefore possible to determine the configurational stability at tin by looking at the NMR spectrum of a racemic mixture of suitably substituted organotin compounds of type M (see figure 1) bearing diastereotopic groups. Some of the compounds examined by this elegant method are listed in Table 1.

Analogous conclusions can of course be extracted from the NMR spectra of molecules of type N bearing an asymmetric carbon atom and two diastereotopic methyl groups on tin (see Table 2):

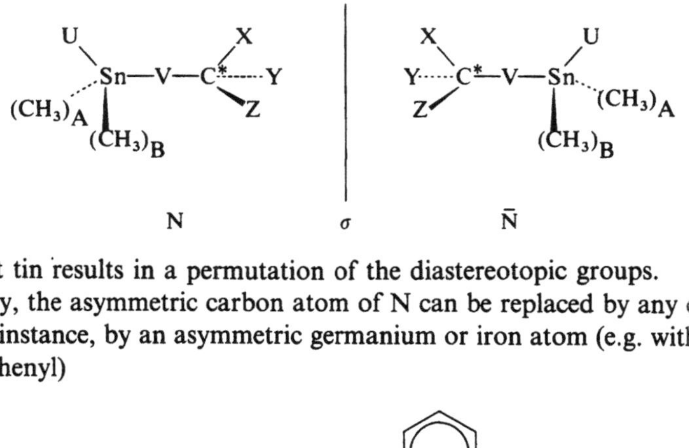

(4)

N σ $\overline{N}$

inversion at tin results in a permutation of the diastereotopic groups.

Obviously, the asymmetric carbon atom of N can be replaced by any other chiral center: for instance, by an asymmetric germanium or iron atom (e.g. with V = nihil and U = phenyl)

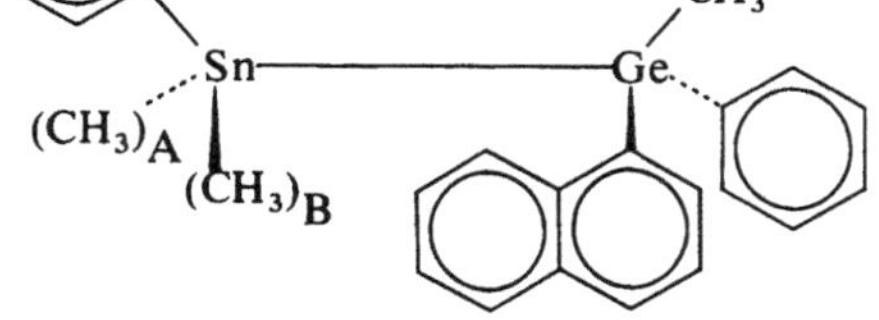

or

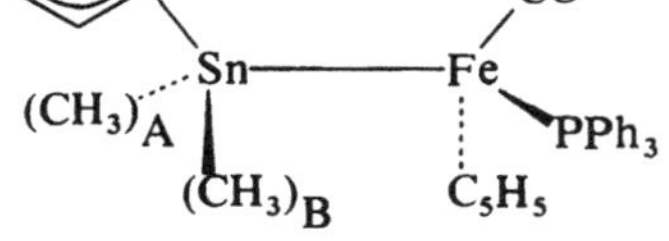

(5)

(19)

Δv_{AB} = 3.2 Hz in CCl_4 [19], 3.5 Hz in HMPA [19] (60 MHz ^{1}H NMR)

(20)

Δv_{AB} = 3.6 Hz (60 MHz ^{1}H NMR); 25 Hz (22.63 MHz ^{13}C NMR) in Pyridine [17]

Table 1. Molecules of type M, i.e. $U(CH_3)_A(CH_3)_BC{-}V{-}SnXYZ$ used to determine the configurational stability at tin. (st. stands for configurationally stable within the NMR time-scale ($\Delta\nu_{AB}$); u., for configurationally unstable; RT stands for room temperature)

Entry	$U(CH_3)_A(CH_3)_BC{-}V{-}SnXYZ$					Solvent	Conclusion	$\Delta\nu_{AB}$	Ref.
	U	V	X	Y	Z		(RT)	(Hz)	
(*2*)	C_6H_5	CH_2	Br	CH_3	C_6H_5	C_6D_6 + Pyridine	u.		[12]
(*3*)	C_6H_5	CH_2	Cl	CH_3	C_6H_5	C_6D_6 + Pyridine	u.		[12]
						CCl_4	st.		[8,10]
						C_6D_6 + DMSO	u.		[8,10]
(*4*)	C_6H_5	CH_2	Br	CH_3	$C(C_6H_5)_3$	o-$Cl_2C_6H_4$	st.	13.2	[13]
						o-$Cl_2C_6H_5$	st. at 150 °C	8.4	[13]
						CCl_4 + 0.5 M DMSO	st.	11.7	[13]
						Pyridine	u.		[13]
(*5*)	C_6H_5	CH_2	Cl	CH_3	$C(C_6H_5)_3$	Pyridine	st.		[23]
(*6*)	C_6H_5	CH_2	Br	CH_3	$C(CH_3)_3$	CCl_4 + Pyridine (0.1 M)	u.		[13]
						CCl_4 + DMSO (0.05 M)	u.		[13]
(*7*)	H	—	Br	C_6H_5	2,4,6-$(CH_3)_3C_6H_2$				[11]
(*8*)	C_6H_5	CH_2	OC_6H_5	CH_3	C_6H_5	C_6D_6 + DMSO	u.		[11]
(*9*)	H	—	CH_3	$CH(CH_3)_2$	C_6H_5	CCl_4	st.		[14]
						CCl_4 + 7 M DMSO	st.		[14]
(*10*)	C_6H_5	CH_2	CH_3	CH_2I	C_6H_5	$CDCl_3$ + Pyridine	st.		[15]
						$CDCl_3$ + DMSO	st.		[15]
(*11*)	C_6H_5	CH_2	CH_3	$C{\equiv}C{-}C_6H_5$	C_6H_5	$CDCl_3$ + Pyridine	st.		[15]
						$CDCl_3$ + DMSO	st.		[15]
(*12*)	C_6H_5	CH_2	H	CH_3	C_6H_5	C_6D_6 + HMPA	st.	4.3	[16]
						DMSO + LiBr	st.	4.3	[16]
(*13*)	C_6H_5	CH_2	$Fe(CO)_2C_5H_5$	CH_3	C_6H_5	CCl_4	st.	2	[17]
						CS_2 + DMSO	st.	2	[17]
						Pyridine	st.	6.7	[17]
							st. at 80 °C	5.3	[17]
						Pyridine	st. (^{13}C)	23	[17]
(*14*)	C_6H_5	CH_2	$Mn(CO)_5$	CH_3	C_6H_5	CS_2/Pyridine 5/1	st.	3.6	[18]
						Pyridine	st. (^{13}C)	22.1	[18]
(*15*)	C_6H_5	CH_2	$Co(CO)_3P(C_6H_5)_3$	CH_3	C_6H_5	Pyridine	st.	10.5	[18]
							st. (^{13}C)	26.5	[18]
(*16*)	C_6H_5	CH_2	$Co(CO)_4$	CH_3	C_6H_5	C_6D_6		3.6	
						C_6D_6 + Pyridine	u.		[18]
(*17*)	C_6H_5	CH_2	$W(CO)_3C_5H_5$	CH_3	C_6H_5		st.		[18]
(*18*)	C_6H_5	CH_2	$Mo(CO)_3C_5H_5$	CH_3	C_6H_5	CD_2Cl_2/DMSO 1/3	st.	9.5	[16]
						$CDCl_2$/Pyridine 1/3	st.	10.0	[16]

Table 2. Molecules of type N, i.e. $U(CH_3)_A(CH_3)_BSn—V—CH(CH_3)Z$ used to determine the configurational stability at tin (st. stands for configurationally stable within the NMR time-scale ($\Delta\nu_{AB}$); u. for configurationally unstable; RT stands for room temperature)

Entry	$U(CH_3)_A(CH_3)_BSn—V—CH(CH_3)Z$			Solvent	Conclusion	$\Delta\nu_{AB}$	Ref.
	U	V	Z		(RT)	(Hz)	
(*25*) *meso*	$CH(CH_3)(CH_2CH_3)$	—	CH_2CH_3	CCl_4	st.	0.9	14)
				CCl_4 + DMSO (7 M)	st.	0.9	14)
(*26*) *meso*	$CH(CH_3)(CH_2CH_2CH_3)$	—	$CH_2CH_2CH_3$	CCl_4	st.		14)
(*27*)	$C(CH_3)_3$	—	CH_2CH_3	CCl_4	st.	0.5	21)
(*28*) *erythro*	$CH(CH_3)(CH_2CH_2CH_3)$	—	CH_2CH_3	CCl_4	st.	0.6	21)
(*29*)	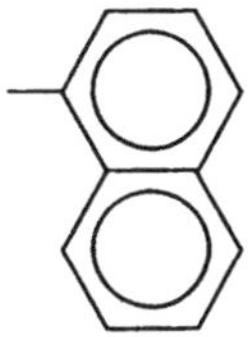	—	CH_2CH_3	CCl_4	st.	0.7	21)
(*30*)	$C(CH_3)_2CH_2CH_3$	—	CH_2CH_3	CCl_4	st.	0.5	21)
(*31*)	CH_2CH_3	—	C_6H_5	CCl_4	st.	2.1	21)
(*32*)	$CH_2CH_2CH_3$	—	C_6H_5	CCl_4	st.	2.4	21)
(*33*)	$CH_2CH_2CH(CH_3)_2$	—	C_6H_5	CCl_4	st.	2.5	21)
(*34*)	$CH_2CH_2C_6H_5$	—	C_6H_5	CCl_4	st.	2.8	21)
(*35*)	$CH_2CH(CH_3)_2$	—	C_6H_5	CCl_4	st.	2.8	21)
(*36*)	$CH_2C(CH_3)_2C_6H_5$	—	C_6H_5	CCl_4	st.	3.1	21)
(*37*)	$CH(CH_3)_2$	—	C_6H_5	CCl_4	st.	4.9	21)
(*38*)	$C(CH_3)_3$	—	C_6H_5	CCl_4	st.	9.2	21)
				CH_3OH	st.	10.0	21)
				Pyridine	st.	8.3	21)
				DMSO	st.	9.9	21)
				HMPA	st.	9.8	21)

Table 2 (continued)

Entry	$U(CH_3)_A(CH_3)_BSn{-}V{-}CH(CH_3)Z$			Solvent	Conclusion	$\Delta\nu_{AB}$	Ref.
	U	V	Z		(RT)	(Hz)	
(39)	C_6H_5	—	C_6H_5	CCl_4	st.	1.6	[21]
				CH_3OH	st.	1.7	[21]
				Pyridine	st.	1.5	[21]
				DMSO	st.	1.8	[21]
				HMPA	st.	1.8	[21]
(40)	C_6H_5	CH_2	C_6H_5	DMSO: dioxane 1:2	st. at 160°	3.7	[8,10]
(41)	Cl	CH_2	C_6H_5	CCl_4	st.	10.3	[22]
				CCl_4 + DMSO	u.		[22]
(42)	Br	CH_2	C_6H_5	CCl_4	st.	7.3	[22]
				CCl_4 + DMSO	u.		[22]
(43)	H	CH_2	C_6H_5	C_6D_6	st.	1.3	[22]
(44)	$N(CH_2CH_3)_2$	CH_2	C_6H_5	C_6D_6	st.	19.2	[22]
				C_6D_6 + HMPA (2.3 M)	st.	19.2	[22]
(45)	$P(C_6H_5)_2$	CH_2	C_6H_5	C_6D_6 + HMPA (3 M)	st.	14.7	[22]
(46)	$As(C_6H_5)_2$	CH_2	C_6H_5	C_6D_6 + HMPA (3.5 M)	st.	11.1	[22]

Even two methyl groups on a tetracoordinate nitrogen atom can be used as a probe to determine the configurational stability at the tin atom. If the nitrogen atom is one of the ligands of the tin atom, then the coalescence of these two signals can be due either to an inversion at tin, nitrogen remaining tetracoordinate or to a Sn—N dissociation. The presence of other diastereotopic groups within the same molecule can serve to distinguish between these two possibilities, as in compound *(21)* [24]

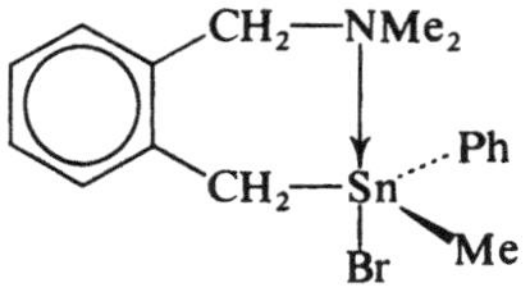

(6)

(21)

for which the diastereotopic methyl groups ($\Delta\nu$ = 38 Hz) coalesce at —37 °C; the diastereotopic protons of the CH_2Sn ($\Delta\nu$ = 15 Hz) and of the CH_2N ($\Delta\nu$ = 9 Hz) methylene groups coalesce at a higher temperature (around 0 °C). This compound is therefore configurationally unstable at the tin atom within the NMR time-scale. On the contrary, compound *(22)* [25]

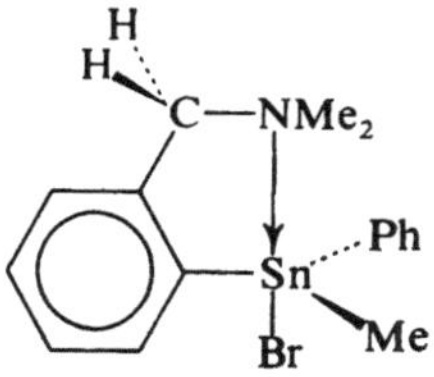

(7)

(22)

is configurationally stable at the tin atom within the NMR time-scale up to 123 °C [25]: the methylene protons show the expected AB pattern even at this temperature ($\Delta\nu$ = 15.5 Hz at 123 °C and 21 Hz at 0 °C) (see also Ref. [88]); even if the two singlets ($\Delta\nu$ = 22 Hz) observed below 30 °C for the CH_3N groups coalesce above 30 °C in the absence or in the presence of pyridine, showing that Sn—N bond dissociation occurs within the NMR time-scale but that the tin atom does not invert at the same rate.

An analogous behavior is observed for compound *(23)* [26]:

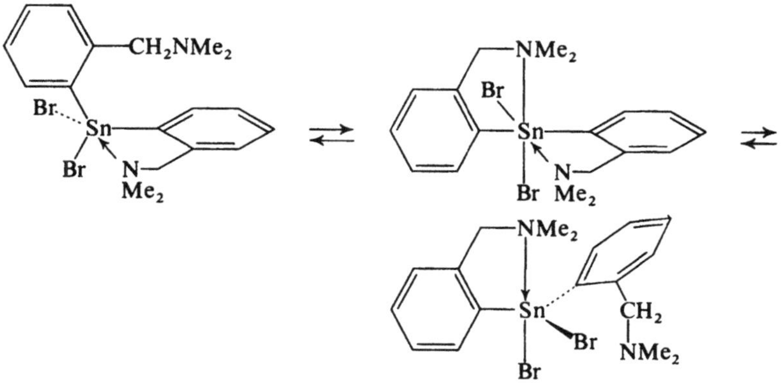

(8)

the two singlets observed at low temperature for the methyls on N coalesce at −50 °C into one singlet whereas the AB pattern observed for the methylene group is retained up to +50 °C. Here again, the NMe groups become homotopic as a result of the nitrogen-tin bond dissociation but the asymmetry at tin is not lost during this process.

If two asymmetric atoms are present in the molecule (see 4.1), NMR can still be used to determine the configurational stability but intermolecularly diastereotopic groups are then used instead of intramolecularly diastereotopic ones.

So, (α-methylbenzyl)cyclohexylisopropylmethyltin (*24*) can exist as two diastereomeric racemic mixtures [21].

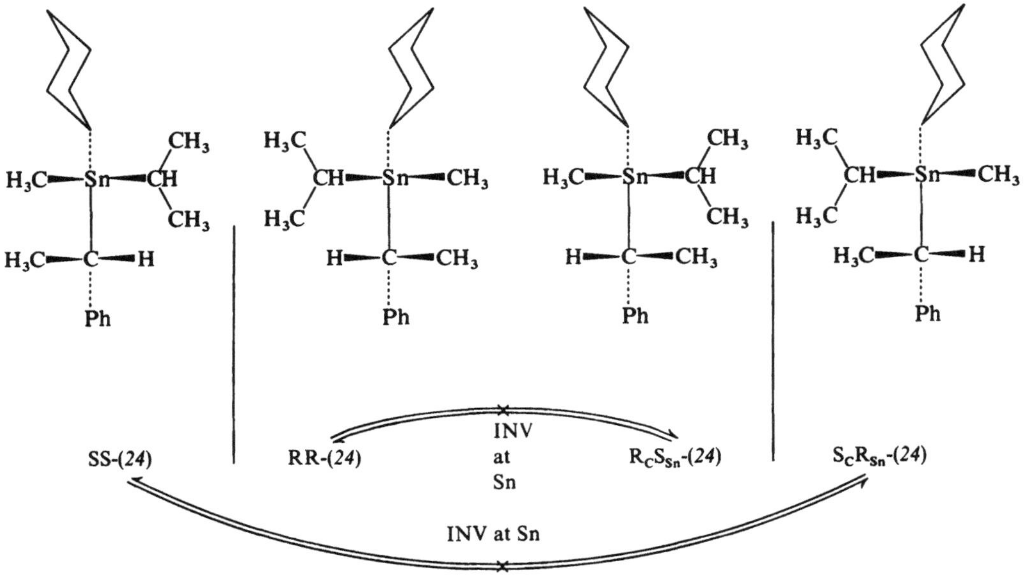

(9)

The two CH_3Sn groups of SS and RR are enantiotopic and give one signal in an achiral medium. The two CH_3Sn groups of RS and SR are enantiotopic too; they give another signal because SS (RR) and RS (SR) are diastereomers. The fact that no coalescence of these two signals is observed [21] shows that compound (*24*) is configurationally stable within the NMR time-scale because the inversion at tin would transform SS into SR and RR into RS.

A triorganotin halide, bis(2-butyl)methyltin chloride [14], with more than one asymmetric atom has been used to show the configurational instability of such compounds in the presence of nucleophiles. It can exist as four different isomers.

They are described below (R or S are the absolute configurations of the 2-butyl groups):

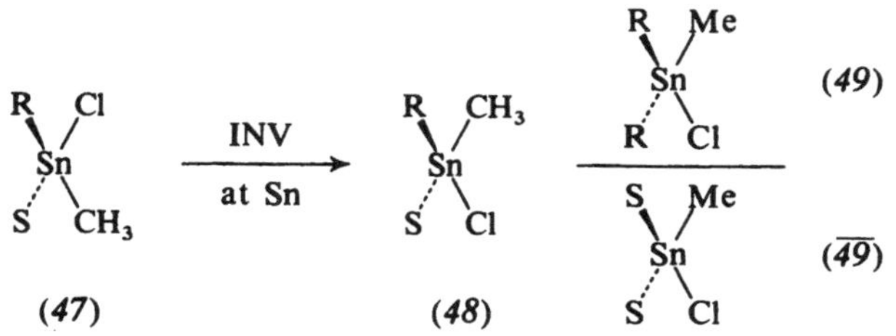
(10)

compounds (*49*) and ($\overline{49}$) are enantiomers whereas (*47*) and (*48*) are achiral compounds (the CH_3SnCl plane is a symmetry plane for both molecules) and are diastereomers; (*47*) is transformed into (*48*) if tin inverts its configuration. The NMR spectrum of (*47*) + (*48*) + (*49*) + ($\overline{49}$) in CS_2 shows three CH_3Sn signals, as expected, but in the presence of very small amounts of DMSO (less than 0.03 M) the first and the third of these three peaks [(*47*) and (*48*)] coalesce [14].

It may be mentioned that other interesting molecules from the *static* stereochemical point of view contain the chiral 2-butyl [14,21], 2-pentyl [14,21] or α-methyl benzyl [21].

So, bis(2-butyl)dimethyltin shows three CH_3Sn signals, one for the homotopic methyls of $RRSnMe_2$ and of its enantiomer $SSSnMe_2$, plus two for the diastereotopic CH_3Sn groups, of the meso-compound $RSSn(Me)_A(Me)_B$ [14,21].

Tris(2-butyl)methyltin and tris(2-pentyl)methyltin exist as two diastereomeric couples of antipodes, (RRR + SSS) and (RRS + SSR), which can be separated [14]. The analogous butyltri(indenyl)tin has been studied by ^{13}C NMR [47,48]: the RRR (and SSS) compound contains three homotopic R groups whereas the RRS (and SSR) compounds clearly contain two diasterotopic R(S) groups plus the S(R) group, which is diastereotopic too with respect to the R(S) groups. At —60°, the NMR shows indeed four signals for the C_1 atom of the indenyl groups [47].

Tetraindenyltin is even more interesting: it shows four signals too for the C_1 carbon with intensities of 1:3, 3:1 at —60 °C [48]. It can exist as five isomers: the RRRR compound (and its enantiomer SSSS) contains four homotopic groups. They must give only one signal (intensity: 4 + 4 = 8) [48]. The RRRS compound (like the SSSR one) contains three homotopic R(S) groups and a S(R) one which is diastereotopic with respect to the three other ones. Statistically, because the energies of the different possible compounds are very similar indeed [48], RRRS, RRSR, RSRR and SRRR (SSSR, SSRS, SRSS and RSSS) are similar possible distributions: therefore the intensities are expected to be $3 \times 8 = 24$ and $1 \times 8 = 8$ [48]. The RRSS is the most exciting molecule: the two homotopic R and the two homotopic S groups can be permuted by the S_4 axis present in this compound: they are therefore enantiotopic and isochronous in achiral medium. RRSS, RSRS, RSSR, SRRS, SRSR and SSRR are the possible distributions. The expected intensity of is therefore $4 \times 6 = 24$ [48].

A diorganotin dihalide with two asymmetric atoms, {2-[1-(S)-dimethylaminoethylphenyl]}methyltin dibromide (*50*), has been described by van Koten [26]. It can exist a priori as two diastereomers, $(S)_C(S)_{Sn}$ and $(S)_C(R)_{Sn}$

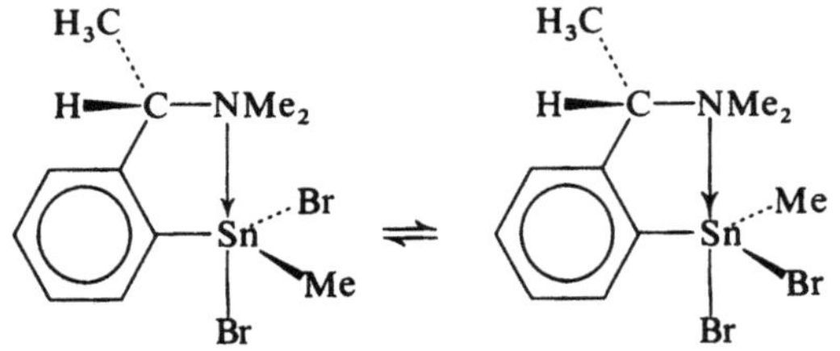

(11)

$(S)_C(R)_{Sn}$-(*50*) $(S)_C(S)_{Sn}$-(*50*)

for which four CH_3N signals are expected. Only two CH_3N signals are observed at room temperature. Therefore, it must be concluded that the nitrogen ligand coordinates to tin: a planar dimethylamino group would give only one signal (see below: coalescence at higher temperatures) and that rapid inversion of configuration at tin occurs. At 58 °C these two methyl signals coalesce to become a single line at 105 °C.

The configurational stability of tin in a monoorganotin complex has been determined by NMR too [33]: compounds of the type $Y{-}CH_2Sn(acac)_2X$ (Y = I, $CH = CH_2$, COOEt) show four acetylacetonate methyl signals (and two acetylacetonate CH signals) below −100 °C (in accordance with a cis configuration) which broaden and collapse between 0° and 40 °C because of an intramolecular averaging process.

The NMR spectrum of the *threo* + *erythro* mixture of methylphenyl(2-phenylpropyl)tin hydride (*51*), another compound with two asymmetric atoms, shows two CH_3Sn signals ($\Delta\nu$ = 6.6 Hz in benzene at 270 MHz) which remain anisochronous even in the presence of HMPA [22]. The analogous *erythro* + *threo* triorganotin deuteride (*52*) shows two CH_3Sn signals too ($\Delta\nu$ = 1.5 Hz at 60 MHz in C_6D_6 [22]).

Tables 1 and 2 show that *triorganotin halides* are configurationally unstable within the NMR time-scale in the presence of traces of nucleophiles such as pyridine or DMSO for instance (entries *2*, *3* and *6* of table 1; entries *41* and *42* of Table 2); however, the presence of bulky groups at tin (like a trityl group for instance) increases their configurational stability (entries *4* and *5* of Table 1); *triorganotin phenoxides* are configurationally unstable within the NMR time-scale in the presence of DMSO (entry *8* of Table 1); *tetraorganotin compounds* are configurationally stable within the NMR time-scale even in the presence of large quantities of nucleophiles (entries *9–11* of Table 1; entries *25–40* of Table 2); *triorganotin hydrides* are configurationally stable within the NMR time-scale even in the presence of strong nucleophiles like HMPA (entry *12* of Table 1; entry *43* of Table 2); *triorganostannyl-molybdenum, -iron and -manganese complexes* are configurationally stable within the NMR time scale even in the presence of strong nucleophiles (entries *18*, *13* and *14* of Table 1); the *triorganostannyltetracarbonylcobalt* (*16*) is configurationally unstable within the NMR time scale in the presence of pyridine, whereas the triorganostannyltricarbonyl(triphenylphosphine)cobalt (*15*) is configurationally stable within the NMR time-scale even in pyridine; *triorganostannylamines, -phosphines and -arsines* are configurationally stable within the NMR time-scale, even in the presence of HMPA (entries *44–46* of Table 2).

2.2 Possible Mechanisms Explaining the Configurational Instability of Triorganotin Halides

Nucleophiles do cause the coalescence of the signals due to diastereotopic groups of the triorganotin halides mentioned in section 2.1. Analogously, triorganosilicon halides (see below) and methyl-1-naphthylphenylgermanium chloride are optically stable in hydrocarbons, in CCl_4 or in $CHCl_3$, but racemize rapidly in THF [28].

In order to get more information about the mechanism explaining the configurational instability of these molecules, the order with respect to the nucleophile was determined by NMR. As seen before, the inversion at tin occurs at a rate v which can be expressed as

$$v/[M] = (\pi/\sqrt{2})\,\Delta v_{\infty} \text{ (in Herz)} \qquad \text{(see 2.1)}$$

On the other hand, v/[M] can be expressed as a function of the concentration of the nucleophile

$$v/[M] = k_2[N] + k_3[N]^2$$

Knowing that Δv_{∞} (in Herz) = $\Delta\sigma_{\infty}$ (in ppm) × Ho (in MHz) and assuming that $\Delta\sigma_{\infty}$ is practically identical in C_6D_6 and in C_6D_6 containing less than 3% pyridine [12] we get

$$Ho \times \Delta\sigma_{\infty} \times \pi/\sqrt{2} = k_2[N] + k_3[N]^2$$

This can be rewritten as

$$Ho/[N] = (\sqrt{2}/\pi\,\Delta\sigma_{\infty})\,(k_2 + k_3[N])$$

The determination of the concentrations in pyridine [N] which cause the coalescence of the signals of the diastereotopic groups of a 0.262 M solution methylneophyl-*t*-butyltin bromide (*6*) and of 0.332 M solution methylneophylphenyltin chloride (*3*) at 22 °C at respectively 60, 100 and 270 MHz shows [12] that the k_2 term is much smaller than the $k_3[N]$ term. From these results, it is clear that the inversion of the configuration of the metal atom of triorganotin halides is second-order in the nucleophile pyridine. An analogous rate equation has been found for the racemization of triorganosilicon halides [29], for which the activation entropy $\Delta S^{\neq}$ is about −50 e.u. Several mechanisms with increase of coordination number [30] can be proposed to account for this second order in the nucleophile [31]:

a) a S_N2 at the metal atom of the triorganotin halide by pyridine (N), followed by the addition of a second pyridine molecule at the tin atom to give an achiral adduct which can give back either the R-triorganotin halide or the S one (see Fig. 2)

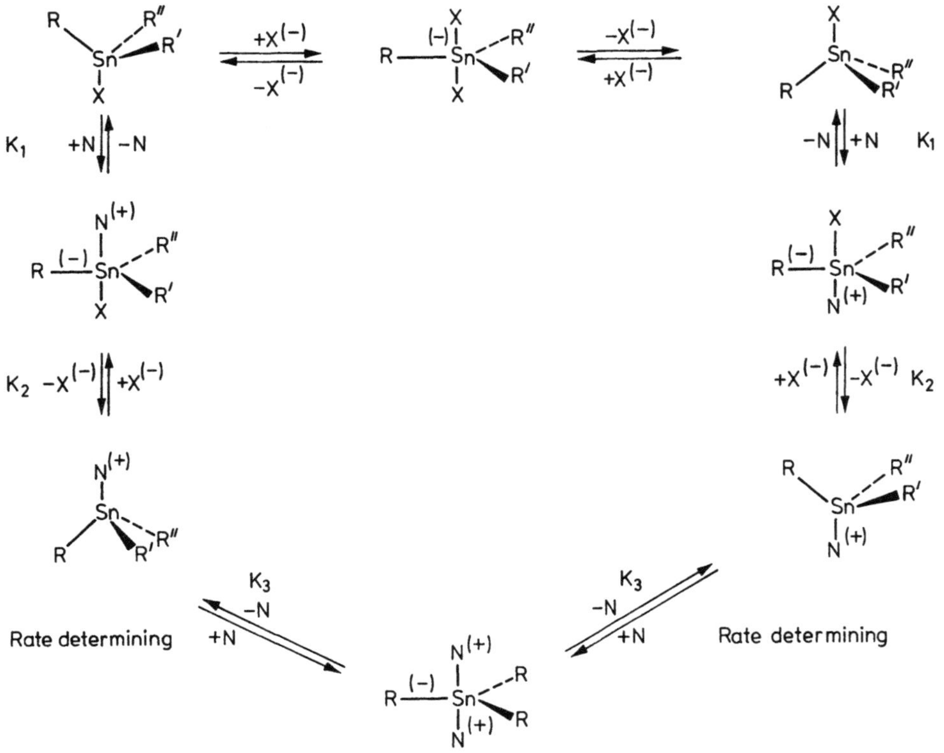

Fig. 2. Possible mechanism for the optical instability of triorganotin halides catalyzed by the nucleophile N

b) the successive additions of two pyridine molecules at the tin atom without loss of halide to give a six-coordinate achiral adduct which can loose the two pyridine molecules to give back either the R-triorganotin halide or the S one.

The first step of both mechanisms is the same, namely the addition of pyridine at the electrophilic metal atom of the triorganotin halide to give a pentacoordinate adduct.

Mechanism a) operates if the tin-halogen bond of this trigonal bipyramidal complex is broken.

Mechanism b) applies if a second solvent molecule is added at the tin atom of this pentacoordinate complex. The influence of the halogen atom can safely be predicated if mechanism a) is operative. The equilibrium constant K_1 for the formation of the trigonal bipyramidal complex is not very different for X=Cl and for X=Br [31]. Furthermore, since $Cl^{(-)}$ is harder than $Br^{(-)}$, the cleavage of the tin-halogen bond of 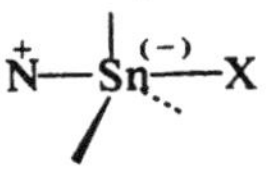 to give 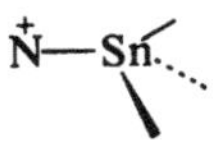

must occur more easily for the bromide than for the chloride, since Sn(IV) is hard. Therefore the triorganotin bromide should racemize more rapidly than the analogous chloride. Effectively, the diastereotopic signals of methylneophylphenyltin bromide

(*2*) are already a fine single line when 0.05 M or pyridine is present, whereas *coalescence* is observed for methylneophylphenyltin chloride (*3*) in the presence of a 0.1 M concentration in pyridine on the same 60 MHz NMR instrument. Furthermore, the difference in chemical shifts for the diastereotopic groups of (*2*) and (*3*) are identical (5.5 Hz) [31].

The evidence favours mechanism a) involving an ion-pair in which the metal atom is five-coordinate. While these results do not completely rule out mechanism b), they are exactly what would be predicted on the basis of mechanism a) and cannot be easily reconciled with mechanism b).

The influence of substituents on the phenyl bound to tin on the configurational stability of methylneophylphenyltin chloride (*3*) has been studied [49]. Methylneophyl-*p*-trifluoromethylphenyltin chloride (*69*) is less optically stable than compound (*3*) [15]. On the contrary, a *p*-trifluoromethyl group totally inhibits the racemization of ethyl-1-naphthylphenylsilicon chloride [32].

The influence of bulky groups on tin on the configurational stability of triorganotin halides can best be explained by their steric hindrance, rendering the tin atom less available for nucleophiles and therefore more configurationally stable. Similarly, the fact that compound (*22*) is configurationally stable within the NMR time-scale even in the presence of pyridine, whereas compounds (*3*) and (*6*) are not, can be ascribed to the possibility of intramolecular coordination in (*22*) which strongly competes with intermolecular coordination necessary to go to the achiral complex. However an analogous intramolecular coordination [giving a six-membered ring instead of a five-membered one, as for (*22*)] is possible for compound (*21*) which is however much less configurationally stable than (*22*).

3 Methods of Synthesis of Optically Active Organotin Compounds

Several methods can be used to prepare optically active organotin compounds in which the tin atom is the only chiral center.

The classical method, which was followed to prepare the first example of an optically pure chiral organotin compound, is characterized by the use of a *auxiliary chiral group* necessary to convert the racemic mixture of enantiomers into a mixture of diastereomers which are then separated by a suitable physical method and converted back into the separated enantiomers by splitting off the chiral auxiliary group. This last step is sometimes difficult to achieve [34].

A second method is the replacement of a *chiral leaving group* of an (optically unstable) organotin compound (a triorganotin menthoxide for instance) by a more nucleophilic reagent (a Grignard reagent or lithium aluminum hydride for instance).

Another asymmetric synthesis is the substitution of a good achiral leaving group by reaction with a *chiral reagent*.

A fourth method is a *chromatographic resolution* of a racemic mixture of organotin compounds for instance on a chiral matrix such as microcrystalline cellulose triacetate.

A fifth and last method is a *stereoselective or -specific substitution reaction* on an optically active organotin compound prepared by one of the four former methods. This last method will be discussed in Section 5.

3.1 Separation of Diastereomeric Organotin Compounds Followed by the Cleavage of the Auxiliary Chiral Group

Although optically active organosilicon and -germanium compounds have been readily available for twenty years [35], the first example of a chiral organotin compound, (+)-3-(*p*-anisylmethyl-1-naphthylstannyl)-1,1-dimethyl-1-propanol, (+)-(*53*), is less than ten years old [36]. Optically active organolead compounds are not yet known, even though racemic organolead compounds are now readily available [37].

Compound (+)-(*53*) has been made from one of the diastereomers of the (—)-menthyl ester of 3-(*p*-anisylmethyl-1-naphthylstannyl)propionic acid, (*54*) ($[\alpha]_D^{20^\circ}$ — 24) which could be obtained from the mixture of diastereomers because it is much less soluble in *n*-pentane at low temperature than the other one. Their separation could be followed by NMR, both diastereomers differing by the position of their methoxy signal. The pure less soluble diastereomer (*54*) reacts with methylmagnesium iodide to give a tetraorganotin compound containing only one chiral center, the asymmetric tin atom [36,87].

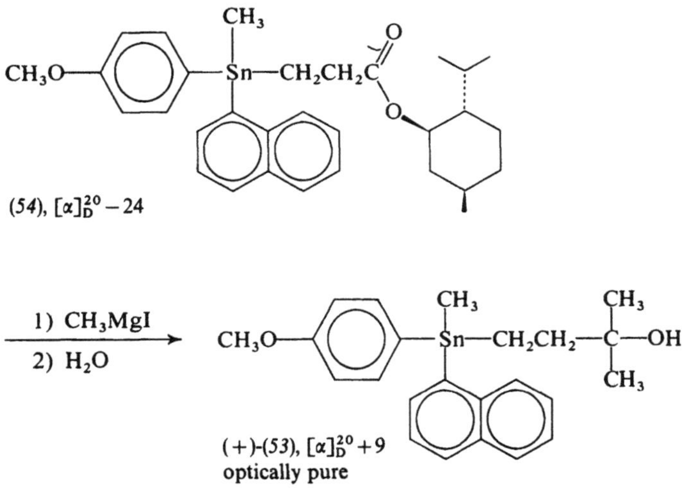

(*54*), $[\alpha]_D^{20}$ −24

(+)-(*53*), $[\alpha]_D^{20}$ +9
optically pure

(12)

It may be mentioned that the diastereomers of the analogous compound bearing a hydrogen atom instead of the methoxy group, the (—)-methyl esters of 3-(methyl-1-naphthylphenylstannyl)propionic acid, could not be separated analogously. Moreover, NMR spectroscopy could not be used as in the former case to determine the diastereomeric ratio.

Optically active (+)- and (—)-*p*-(*i*-propylmethylphenylstannyl) benzoic acids (*56*) and their methyl esters (*57*) were similarly prepared by Lequan four years later [38] (see Table 3). They are characterized by very low optical rotations. Furthermore, the diastereomeric brucine salts via which the acids were resolved, are characterized by almost identical NMR spectra that cannot be used to follow their separation so that no precise information is available about the optical purity of (*56*) and (*57*).

An analogous synthesis, via its menthydrazone (*59*), of optically active (+)-*p*-(cyclohexylmethyl-1-naphthylstannyl) acetophenone, (+)-(*58*), which is characterized

Table 3. Synthesis of optically active organotin compounds RR'R"S$\overset{*}{n}$L' from one of the diastereomeric compounds bearing the auxiliary chiral group R*: RR'R"S$\overset{*}{n}$LR* → RR'R"S$\overset{*}{n}$—L' (*: optically pure; **: optical purity unknown)

R	R'	R"	Ligand with the auxiliary chiral group LR*	$[\alpha]_D$ of RR'R"S$\overset{*}{n}$LR*	Reagent cleaving the auxiliary chiral group	L'	$[\alpha]_D$ of RR'R"S$\overset{*}{n}$L'	Ref.
CH_3	p-CH_3O—C_6H_4	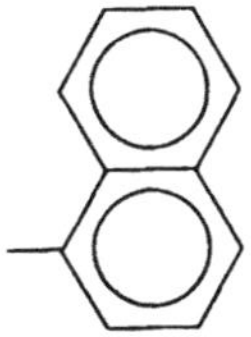	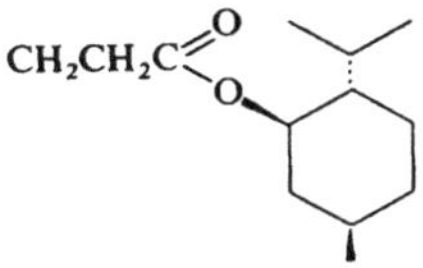	(*54*) —24	CH_3MgI	$CH_2CH_2C(CH_3)_2OH$	(*53*) +9	36)*
CH_3	$(CH_3)_2CH$	C_6H_5	C_6H_4—p—$C(=O)O^{(-)}$ [(—)-Brucine H]$^{(+)}$	—57	HCl	C_6H_4–p–$C(=O)OH$	(*56*) —0.5	38)**
						↓ CH_2 C_6H_4–p–$C(=O)OCH_3$	(*57*) —0.1	38)**
CH_3	$(CH_3)_2CH$	$C_6H_5CH_2$	C_6H_4—p—$C(=O)O^{(-)}$ [(—)-Brucine H]$^{(+)}$	—28.1	HCl	C_6H_4–p–$C(=O)OH$	(*55*) —0.6	38)**
CH_3	CH_2<CH_2–CH_2/CH_2–CH_2>CH	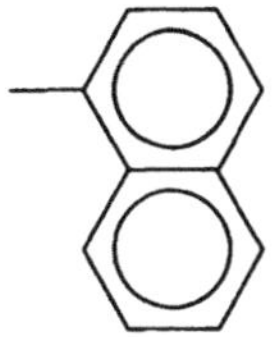	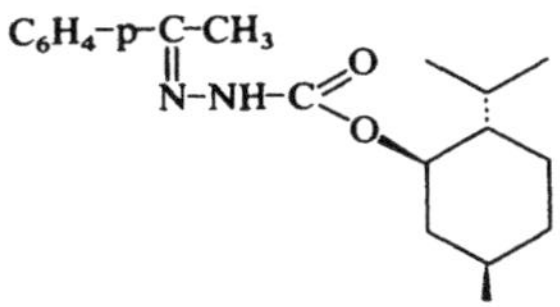	(*59*) —32.7	CH_3COCH_3 TsOH	C_6H_4–p–$C(=O)$–CH_3	(*58*) $[\alpha]_{436}$ = +0.25	39)**
CH_3	$(CH_3)_2CH$	C_6H_5	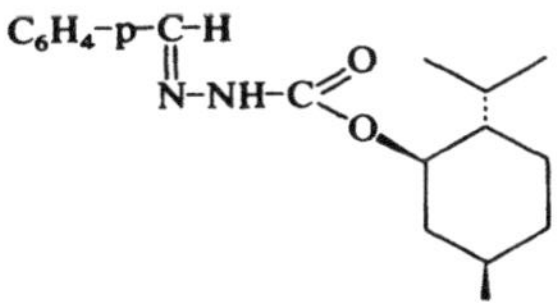	(*61*) —34.6; $[\alpha]_{436}$ = —71.4	CH_3COCH_3 TsOH	C_6H_4–p–$C(=O)$–H	(*60*) $[\alpha]_{436}$ = +0.35	40)**

by a quite low optical rotation too, was reported in 1977 [39] (see Table 3). Here again, no information about the optical purity of (*58*) could be obtained because two different mixtures of diastereomeric (*59*) show almost the same NMR spectrum.

A similar preparation of (+)-*p*-(*i*-propylmethylphenylstannyl)benzaldehyde (*60*) of unknown optical purity was described three years later by Lequan [40] (see Table 3).

The problems associated with the use of this classical method in *organotin* chemistry are essentially due to the fact that the carbon-tin bond can sometimes very easily be cleaved by electrophiles or by nucleophiles. The crucial step is therefore the elimination of the auxiliary group without the cleavage of any of the carbon-tin bonds. This cleavage could for instance not be achieved successfully in the case of *p*-(*i*-propylmethylphenylstannyl)-N,N-dimethylaniline [formula (*60*) in which 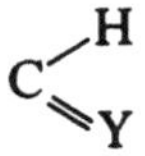 is replaced by $N(CH_3)_2$] which could not be recovered from its (—)-dibenzoyltartrate [34].

Another point which has already been mentioned before, and which is not specific for organotin compounds, is that the diastereomers should be distinguishable by a spectroscopic method (for instance NMR) to be sure that their separation was sufficient enough to give at least one of them in pure form. Till now, this has been the case only once. The methoxy signal seems to be a good probe for this purpose (see above), but unfortunately, the anisyl-tin bond is cleaved much more rapidly than the phenyl- or naphthyl-tin bond, which might sometimes cause problems in the crucial step.

3.2 Replacement of a Chiral Leaving Group by an Achiral Nucleophile

In 1973, Taddei succeeded in preparing optically active (+)-benzyl-*i*-propylmethylphenyltin, (+)-(*62*), by a reaction of the optically unstable *i*-propylmethylphenyltin menthoxide with benzylmagnesium bromide [11] (see Table 4). This asymmetric induction very probably does *not* yield an optically pure compound but is a very rapid and facile route to optically active organotin compounds of unknown optical purity.

It may be mentioned that the optical yield of the analogous reaction between methylneophyltrityltin menthoxide and *i*-propylmagnesium bromide is different at —15 °C and at 0 °C: at —15 °C, optically active (+)-*i*-propylmethylneophyltrityltin, (+)-(*63*), is obtained (see Table 4) whereas at 0 °C optically inactive (*63*) is obtained [41].

Other chiral leaving groups have been used (see Table 4), for instance (—)-methyl-S-thioglycolate [42], or the conjugate base of cinchonine [92], which seems to give rather good results yielding for instance benzylmethylphenyl-t-butyltin with $[\alpha]^{20}_{436} = -22.6$.

The menthoxide ion can be displaced by other nucleophiles, for instance by the triphenylgermyl anion: *i*-propylmethylphenyltin menthoxide reacts with triphenylgermyllithium to give optically active (+)-*i*-propylmethylphenyl(triphenylgermyl)tin, (*64*), of unknown optical purity [19] (see Table 4).

A hydride can displace a menthoxide too: the reaction of *t*-butyl- or methyl-

Table 4. Synthesis of optically active organotin compounds RR′R″SnL′ from an optically unstable compound bearing a chiral leaving group Y*

$RR'R''Sn{-}Y^* \xrightarrow{L'-M} RR'R''Sn{-}L'$

R	R′	R″	Y*	L′—M	$[\alpha]_D$ RR′R″SnL′	Ref.
CH_3	$CH(CH_3)_2$	C_6H_5	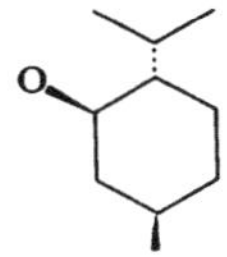	$C_6H_5CH_2{-}MgCl$	*(62)*, +4.6	[11]
CH_3	$CH_2C(CH_3)_2C_6H_5$	$C(C_6H_5)_3$	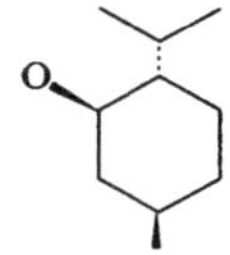	$(CH_3)_2CH{-}MgBr$ (−15 °C)	$[\alpha]_{578} = +0.62$ *(63)*, +5.1	[42] [41]
CH_3	$CH(CH_3)_2$	C_6H_5	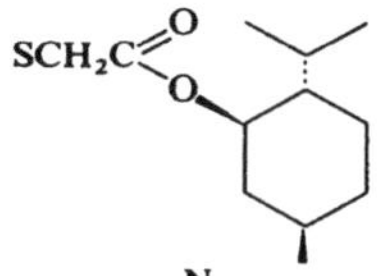	$C_6H_5CH_2{-}MgCl$	*(62)*, $[\alpha]_{578} = +0.83$ $[\alpha]_{436} = +2.1$	[42]
CH_3	$CH(CH_3)_2$	C_6H_5	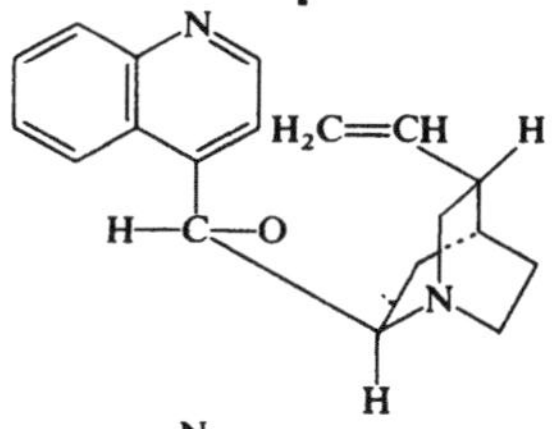	$C_6H_5CH_2{-}MgBr$	$[\alpha]^{20}_{436} = -7$	[92]
CH_3	$CH(CH_3)_2$	C_6H_5	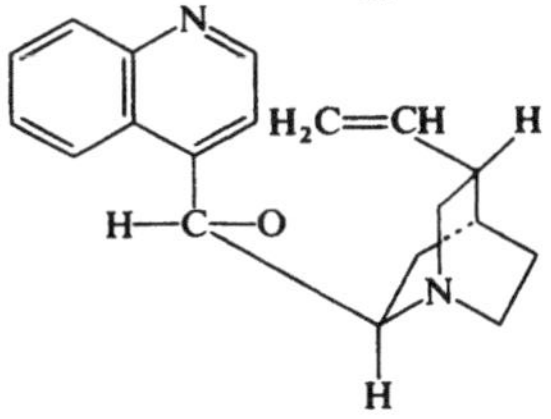	$CH_3CH_2CH_2CH_2{-}Li$	$[\alpha]^{20}_{436} = -1.5$	[92]

Table 4 (continued)

R	R′	R″	Y*	L′—M	$[\alpha]_D$ RR′R″SnL′	Ref.)
CH_3	$CH(CH_3)_2$	C_6H_5	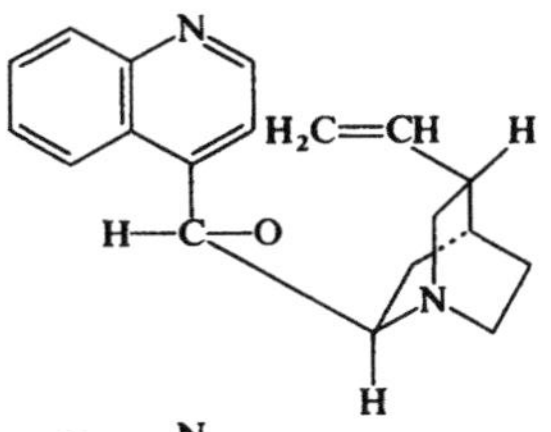	$CH_3CH_2CH_2CH_2—MgX$	$[\alpha]^{20}_{478} = +0.3$	92)
CH_3	$CH(CH_3)_2$	C_6H_5	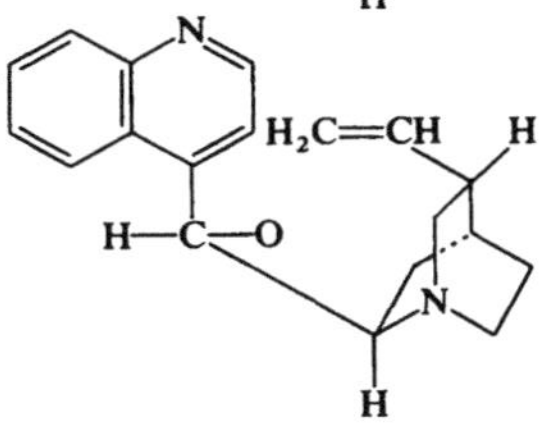		$[\alpha]^{20}_{436} = +1.3$	92)
CH_3	$CH(CH_3)_2$	$C(CH_3)_3$	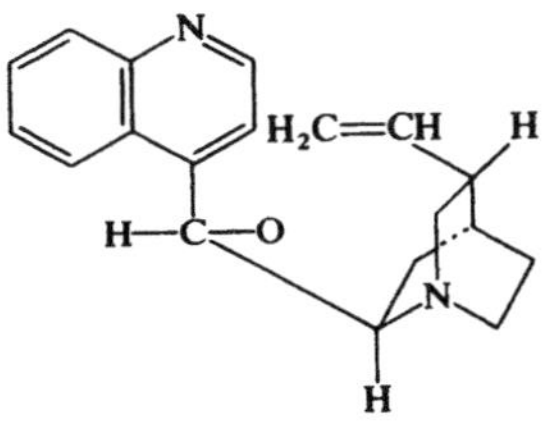	$C_6H_5—MgCl$	$[\alpha]^{20}_{436} = +0.7$	92)
CH_3	$C(CH_3)_3$	C_6H_5	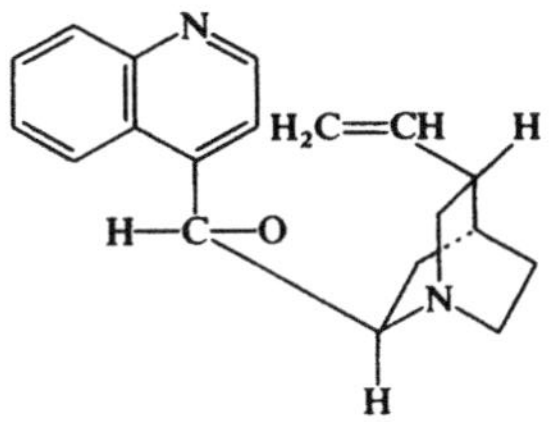	$(CH_3)_2CH—MgBr$	$[\alpha]^{20}_{436} = +4.7$	92)

CH_3	$C(CH_3)_3$	C_6H_5	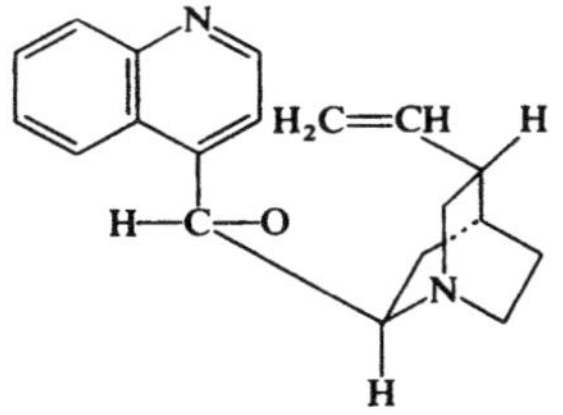	$C_6H_5CH_2MgBr$	$[\alpha]^{20}_{436} = -22.6$	92)
CH_3	$CH(CH_3)_2$	C_6H_5	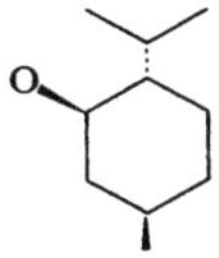	$(C_6H_5)_3Ge{-}Li$	(*64*), $[\alpha]_{436} = +0.7$	19)
$C(CH_3)_3$	$CH_2C(CH_3)_2C_6H_5$	C_6H_5	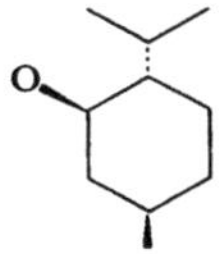	$H{-}AlH_3Li$	(*65*), −0.5	43,44)
CH_3	$CH_2C(CH_3)_2C_6H_5$	C_6H_5	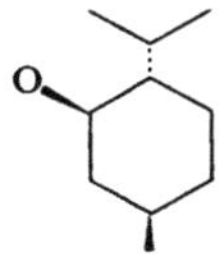	$H{-}AlH_3Li$	(*12*), −0.5	43,44)

Table 5. Synthesis of optically active organotin compounds RR'R''SnL' by the reaction of an optically triorganotin halide with a chiral reagent

$$RR'R''Sn{-}X \xrightarrow{L'-Al(O\text{-}3,5\text{-}Me_2Ph)_2(O\text{-}(-)\text{-}CHPhCHMeNMe_2)} RR'R''Sn{-}L'$$

R	R'	R''	X	L'	$[\alpha]_{365}$ of RR'R''SnL'	Ref.
CH_3	$CH_2C(CH_3)_2C_6H_5$	C_6H_5	Cl	H	(*12*), +6.4	43,44)
$C(CH_3)_3$	$CH_2C(CH_3)_2C_6H_5$	C_6H_5	Cl	H	(*65*), −3.5	43,44)
CH_3	C_6H_5	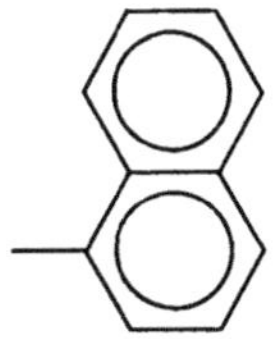	I	H	(*66*), +2.0	44)
CH_3	$CH_2C(CH_3)_2C_6H_5$	C_6H_5	Cl	$C_6H_5{-}C{\equiv}C$	(*11*), $[\alpha]_D$ = −3.8	41,46)

neophylphenyltin menthoxide with lithium aluminum hydride yields the optically active *t*-butyl- or methyl-neophylphenyltin hydride, (*65*) and (*12*), respectively, both of unknown optical purity.

3.3 Substitution of a Good Leaving Group with a Chiral Reagent

The reaction of configurationally unstable triorganotin halides with the chiral reducing agent introduced by Vigneron and Jacquet [45], H-Al(O-3,5-Me_2Ph)$_2$(O-(—)-$\overset{*}{C}$HPh$\overset{*}{C}$HMeNMe_2), yields chiral triorganotin hydrides [20], the optical purities of which seems to be higher than those of the triorganotin hydrides obtained by the former method using a menthoxide ion as leaving group and lithium aluminum hydride as nucleophile (compare Tables 4 and 5). For example, 1-naphthylmethylphenyltin iodide is reduced by this chiral reducing agent into (+)-1-naphthylmethylphenyltin hydride, (+)-(*66*).

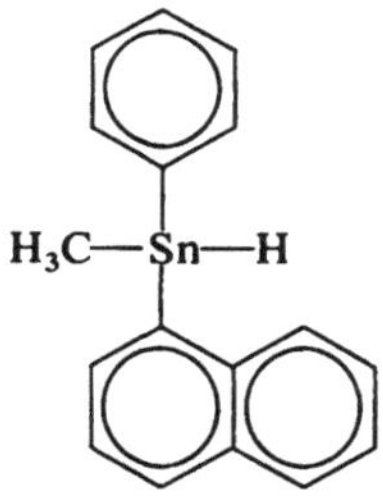

(13)

(*66*)

Analogously, optically active (—)-methylneophylphenyl-(phenylethynyl)tin, (—)-(*11*) has been obtained by the reaction of methylneophylphenyltin chloride (*3*) with the corresponding chiral alkylating agent Ph—C≡C—Al(O-3,5-Me_2Ph)$_2$(O-(—)-CHPhCHMeNMe_2) [41,46].

3.4 Chromatographic Resolution of Racemic Organotin Compounds

The optical purity of almost all the organotin compounds described in this chapter is not yet known. In order to determine the stereoselectivity of substitution reactions at the tin atom of these organotin compounds, it is almost always necessary to know the optical purity of the starting compound and of the final product. The method described in this section can be used not only for the resolution of racemic organotin compounds but also for the determination of their optical purity [50]. It will be a valuable tool for the determination of the stereoselectivity of the reactions described in Chapter 5, and of other reactions which will be studied.

Microcrystalline cellulose triacetate introduced by Hesse and Hagel [51] can be used to resolve organotin compounds [52,53]. Some preliminary results obtained on a rather small column (column A: length: 35 cm; internal diameter: 2 cm; filled with

55 g microcrystalline cellulose triacetate (53μ–105μ) swollen in 95% ethanol; 70 theoretical plates; nitrogen pressure: 300 Torr, elution with 95% ethanol, 200 ml/h; dead volume V_0: 60 ml) are given in Table 6.

A typical diagram, the one obtained for methylneophylphenyltrityltin (*67*) on column A, is given in Fig. 3 [53].

Table 6. Room temperature partial chromatographic resolution of 75 mg tetraorganotin compounds on microcrystalline cellulose triacetate (column A)

Organotin racemate	α_{365} (max) for the first ORD peak	α_{365} (max) for the second	mean $[\alpha]_{365}^{25°}$	Retention Volume V'_R (ml)
(*63*) $Me(PhMe_2CCH_2)(Me_2CH)SnCPh_3$	−0.012°	+0.006°	4.5	15
(*67*) $Me(PhMe_2CCH_2)PhSnCPh_3$	−0.155°	+0.097°	60	30
(*68*) $(PhCH_2)MePhSnCPh_3$	−0.040°	+0.026°	32	50
(*62*) $(PhCH_2)Me(Me_2CH)SnPh$	−0.012°	+0.008°	4.0	40

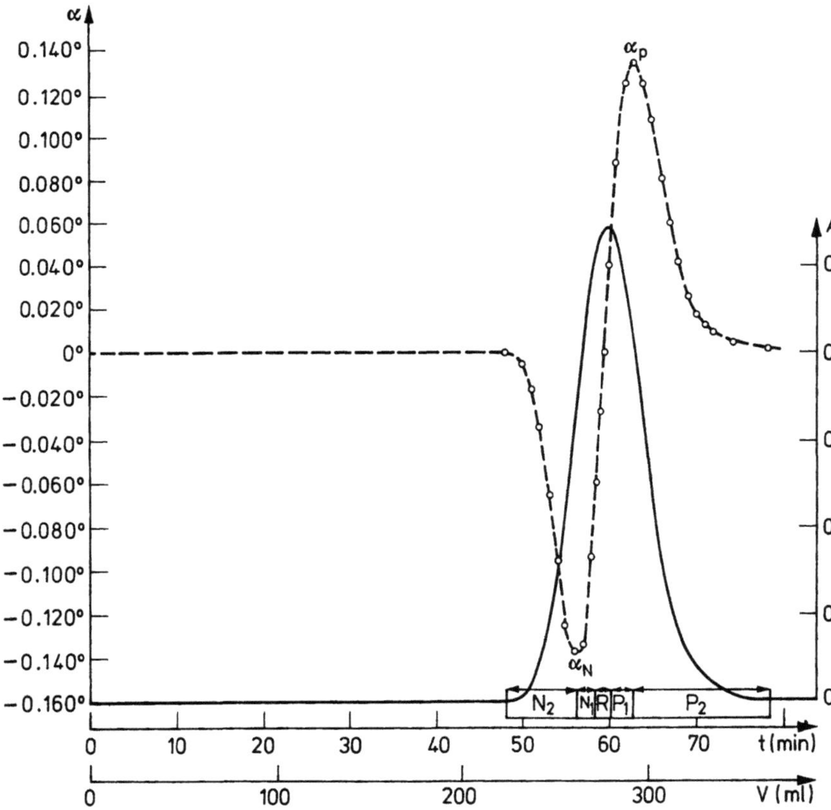

Fig. 3. Partial chromatographic resolution of 75 mg of methylneophylphenyltrityltin (*67*) on microcrystalline cellulose triacetate (column B) [55] (see Table 7). ——— UV (λ = 275 nm); – – – – ORD (λ = 365 nm). (taken from Ref. [55] with permisson)

Table 6 clearly shows that fractions with a high optical rotation have been obtained for at least two organotin compounds.

Already partially resolved methyl-1-naphthylphenylgermanium hydride ($[\alpha]_D^{20°}$ + 10) could also be separated on column A into two fractions characterized by $[\alpha]_{365}^{25°}$ = +80 and —42 respectively. Partial resolution could even be obtained for the configurationally less stable methyl-1-naphthylphenylgermanium chloride if the elution was carried out with diethyl ether [53], which is much less nucleophilic towards germanium than methanol, which causes the racemization of this compound.

This chromatographic method might be used to try the separation of the enantiomers of a racemic mixture of asymmetrically substituted organolead compounds, which are now available [37,54].

Table 7. Influence of temperature on the resolution of 75 mg (*67*) by chromatography on column B [55]. (V'_R = retention volume; t_R: retention time)

T°	0 °C	20 °C	40 °C
α_N (°, 365 nm)	—0.145	—0.138	—0.128
α (°, 365 nm)	+0.132	+0.126	+0.120
$V'_R = V_R - V_o$; t_R (min)	150; 102	156; 60	145; 45
Fractions $[\alpha]_D^{20°}$; quantity (mg)			
N_2	—35.2; 14	—33.3; 15	—30.2; 13
N_1	— 7.5; 16	— 6.4; 19	— 6.8; 20
R	— 3.2; 7	— 0.7; 11	+ 0.2; 10
P_1	+ 5.1; 21	+ 5.6; 16	+ 4.8; 17
P_2	+41.0; 13	+40.1; 13	+32.1; 14

With column B (l = 66 cm, internal φ: 2.8 cm; filled with 135 g microcrystalline cellulose triacetate 43μ–53μ swollen in 95% ethanol; dead volume V_0: 120 ml; maximum allowed external pressure: ca. 780 Torr; 250 ml/h; theoretical number of plates N = 320), racemic (*67*) could be partially resolved too [55]. It is interesting to note that a better resolution is obtained at lower temperatures (see Table 7).

A preparative separation of 1 g racemic (*67*) has been performed at 20 °C by eluting successively 10 samples of 100 mg on column B [55]. The following crystalline fractions were obtained ($[\alpha]_D^{20}$; quantity, in mg): N_2 (—32.2; 220); N_1 (—7.0; 200); R (—0.6; 100); P_1 (+4.0; 250) and P_2 (+39.8; 200). The extreme α-values were α_N = *ca.* —0.270° and α_P = ca. +0.225°. Recrystallization of N_2 or of P_2 from ethanol yielded crystals with a lower, and evaporated mother liquors with a higher optical rotation (see Fig. 4).

The optical purity P of the M_1N_2 fraction was determined [55] (P = 0.62) which would lead to an $[\alpha]_D$ = —209 for pure (—)-(*67*) Chromatography of the mother liquor M_1N_2 on column B gave two fractions M_1N_4 ($[\alpha]_D$ = —155) and M_1N_3 ($[\alpha]_D$ = —100) (see Fig. 4). The first eluted fraction M_1N_4 was recrystallized from a 1:4 MeOH:EtOH mixture. The mother liquor M_2N_4 showed, after evaporation

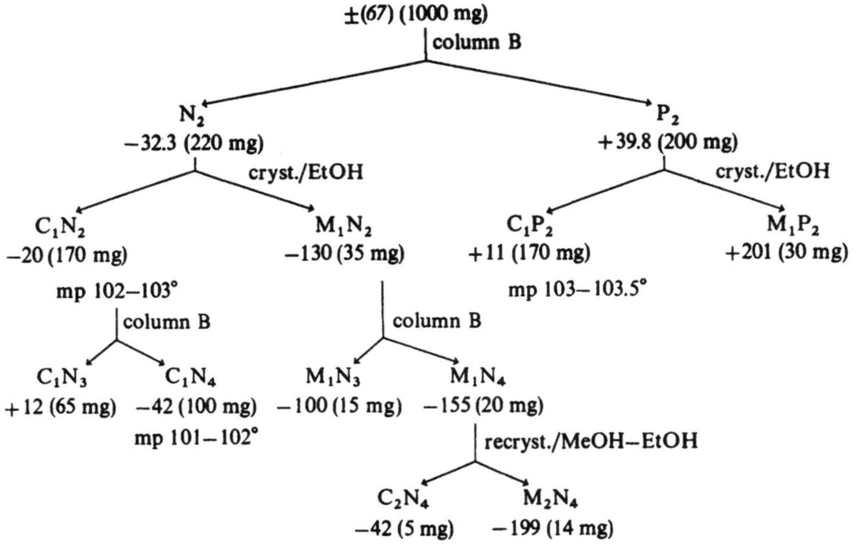

Fig. 4a. Resolution of racemic (*67*) by column chromatography on microcrystalline cellulose triacetate (column B) and by recrystallization [55]; C = crystals; M = mother liquor; the index following C or M gives the number of recrystallizations the fraction has undergone. The value of $[\alpha]_D^{20}$ is given, followed by the quantity obtained

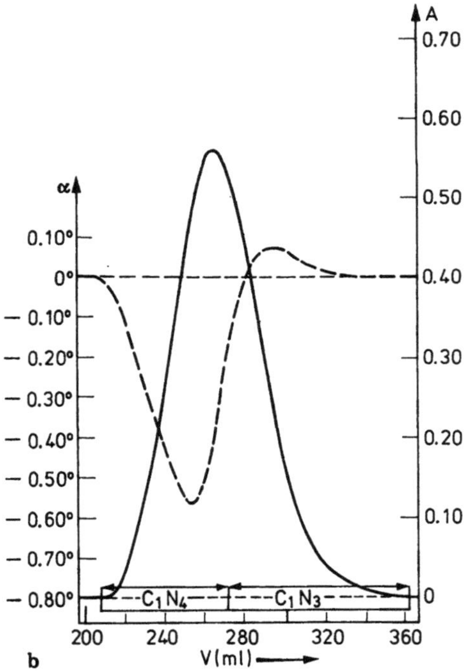

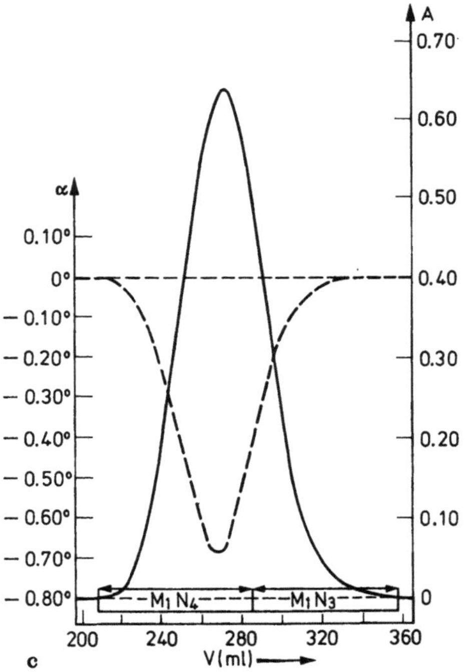

Fig. 4b. Chromatogram of 170 mg MeNe of PhTritSn (*67*), fraction C_1N_2; $[\alpha]_D = -20$. ——— UV ($\lambda = 275$ nm), – – – – α ($\lambda = 365$ nm). (taken from Ref. [55] with permission)

Fig. 4c. Chromatogram of 35 mg MeNeofPhTritSn (*67*), fraction M_1N_2 $[\alpha]_D = -130$. ——— UV ($\lambda = 285$ nm), – – – – α ($\lambda = 365$ nm). (taken from Ref. [55] with permission)

of the solvent, $[\alpha]_D = -199$. Within experimental error, this value is equal to the calculated value for (—)-(*67*). Furthermore, fraction M_1P_2 (see Fig. 4) showed an $[\alpha]_D = +201$. Unfortunately, neither M_1P_2 nor M_2N_4 could be obtained yet in crystalline form.

The influence of the column temperature on the resolution of methylphenyl-*t*-butyl trityltin (*70*) is much more important than for (*67*): at 40 °C, there is no resolution at all on column B, whereas at —10 °C, important α_N and α_P are measured (ca. 0.120°) [55].

It may be noticed that the following compounds could *not* be resolved by chromatography on column B: (*13*), (*15*), (*66*), (*67*), $(PhMeNpSn)_2$ (*71*), $[PhMe(PhMe_2CCH_2)Sn]_2$ (*72*), $PhMe(PhMe_2CCH_2)SnSnPh_3$ (*73*), PhMeNpSnFe-$(CO)_2Cp$ (*74*).

4 Optical Stability of Organotin Compounds

If the coalescence of signals due to diastereotopic groups gives information about the configurational stability of organotin compounds, the information is limited by the NMR time-scale. The configurational stability for longer periods, called from now on optical stability, can, however, be determined by the use of diastereomeric mixtures or of optically active compounds. Thus, optical instability implies that isomeric species interconvert so rapidly that they cannot be separated, whereas configurational instability implies that the positional exchange leading to isomer interconversion is so fast on the observational time-scale (usually the NMR time-scale) that exchanging groups appear to be averaged.

4.1 The Use of Diastereomeric Mixtures

If organotin molecules can be made which contain, besides the chiral tin atom, another asymmetric atom which is optically stable, then they can exist as two diastereomeric racemic mixtures if the other atom is not resolved, or as two diastereomers if the other atom is. If these two diastereomeric mixtures (or diastereomers) can be separated (or at least if two mixtures of different compositions, i.e. diastereomeric ratios, can be obtained), then the rate at which these two mixtures (or compounds) are transformed into the equilibrium mixture is a measure for the rate of inversion at the tin atom.

4.1.1 Optical Stability of Five-Coordinate Triorganotin Complexes

Compound $(S)_{Sn}(S)_C$-(*75*) was obtained [27] by preferential crystallization of the less abundant and less soluble compound of the 40/60 equilibrium mixture of epimerizing $(S)_{Sn}(S)_C$ and $(R)_{Sn}(S)_C$ compounds. Since the two diastereomers have different

NMR spectra [27], the epimerization can be followed easily. At −55 °C, the SS isomer is stable in solution. At −13 °C, epimerization at the tin atom occurs:

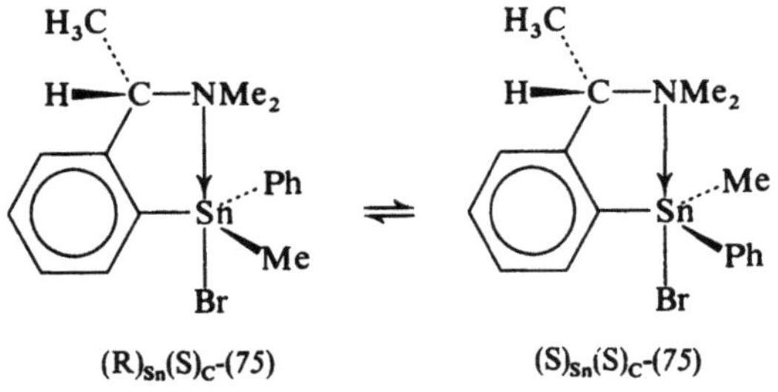

(14)

after $^1/_2$ h. at −13°, a 66/34 mixture is obtained and, after 1.5 h, the equilibrium mixture is reached.

This shows that inversion of configuration at this five-coordinate tin atom *does occur on the laboratory time-scale.*

It is not very clear why van Koten and Noltes [27] did propose the following mechanism

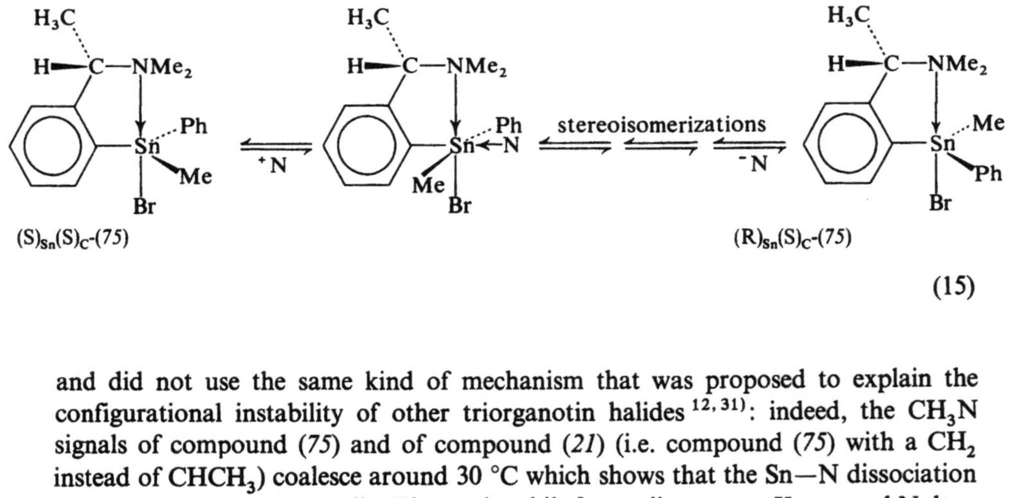

(15)

and did not use the same kind of mechanism that was proposed to explain the configurational instability of other triorganotin halides [12, 31]: indeed, the CH_3N signals of compound (*75*) and of compound (*21*) (i.e. compound (*75*) with a CH_2 instead of $CHCH_3$) coalesce around 30 °C which shows that the Sn—N dissociation process does occur very easily. The nucleophile [according to van Koten and Noltes, another molecule of (*75*)] causing the epimerization of $(S)_{Sn}(S)_C$-(*75*) (see above) might serve as well to add at the tin atom of the four-coordinate species which has been shown to be in equilibrium with five-coordinate (*75*); bromide could be lost and another nucleophile molecule could be added at the tin atom. From this complex containing an achiral tin atom, the reserved reactions could be used to get the $(R)_{Sn}(S)_C$ compound (see Fig. 5). A second order racemization with respect to the nucleophile would support this mechanism and would be sufficient to withdraw van Koten and Noltes' proposal.

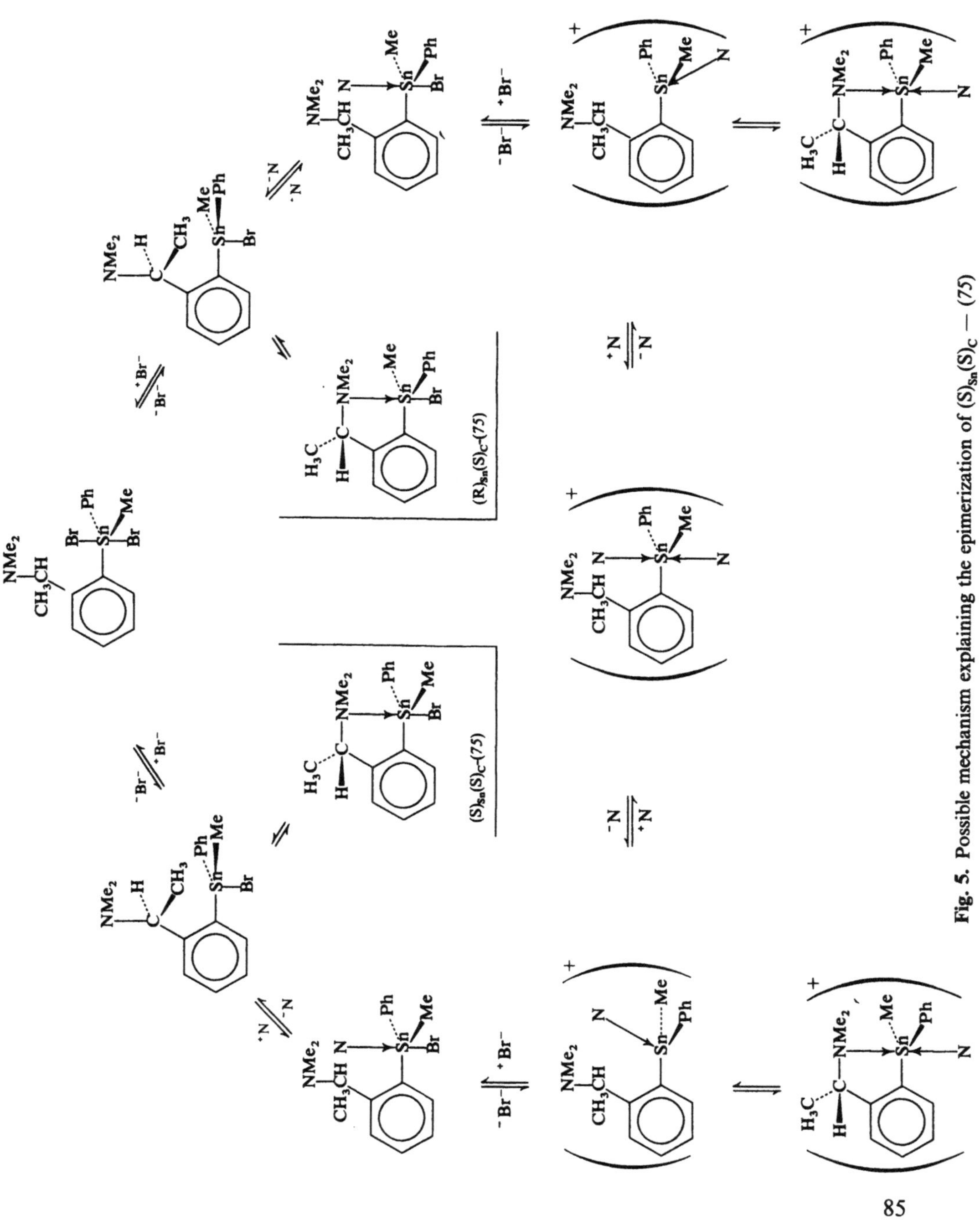

Fig. 5. Possible mechanism explaining the epimerization of (S)$_{Sn}$(S)$_{C}$ — (75)

4.1.2 Optical Stability of Triorganostannyl-Transition Metal Complexes

[Methylphenyl(2-phenylpropyl)stannyl](triphenylphosphine)tricarbonyl cobalt (*76*)

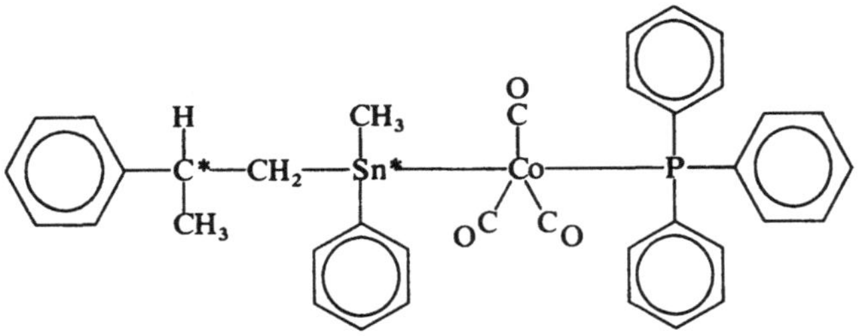

(16)

(*76*)

contains, besides the asymmetric tin atom Sn*, an asymmetric carbon atom C*. It can therefore exist as four stereoisomers: a *threo* and an *erythro* pair of enantiomers (both asymmetric atoms are liked by a methylene and bear two identical ligands, a methyl and a phenyl group [56]).

This mixture is characterized by two anisochronous ^1H_3CSn signals, one for the *threo* and one for the *erythro* mixture. The ^1H_3CSn signal absorbing at higher field is named A, the other one, B. The diastereomeric ratio $(76)_A/(76)_B$ can easily be determined by integration (see Fig. 6).

A standard procedure (see Fig. 9) allows the synthesis of a $(76)_A/(76)_B = 45/55$ mixture [18], which is a crystalline solid which can be enriched in one of the diastereomers by fractional recrystallization [18, 57] (see Fig. 7).

The composition of any of the mixtures obtained remains unchanged after melting under nitrogen (between 90 °C and 105 °C), or after standing at room temperature for several months. Almost pure $(76)_A$ was obtained after twenty recrystallizations. An X-ray diffraction study has shown [57] that it is the *threo*-mixture (see Fig. 8).

This shows that compounds (*76*) is optically stable at tin, i.e. that tin does not invert even after long periods. However, pyridine causes a slow epimerization. This was an example of the case where the asymmetric carbon atom was not resolved and where two diastereomeric racemic mixtures could be separated.

Other methylphenyl(2-phenylpropyl)stannyl-transition metal complexes are oily compounds [18]. Fractional recrystallization could therefore not be applied to separate those diastereomers. For the irondicarbonylcyclopentadienyl compound (*77*), the diastereomeric ratio $(77)_A/(77)_B = 45/55$ could be reached by the standard route (see Fig. 9) but could not be modified by column chromatography.

However, another reaction route to (*77*) gave a $(77)_A:(77)_B$ ratio equal to 60:40 [18].

$$Ph_2MeSnCl \rightarrow Ph_2MeSnFe(CO)_2Cp \xrightarrow{HCl} PhMeClSnFe(CO)_2Cp \xrightarrow{R^*MgCl}$$
$$PhMeR^*SnFe(CO)_2Cp \quad (77)_A:(77)_B = 60:40$$

The composition of these two different diastereomeric mixtures remain unchanged for weeks in pyridine. This shows that (*77*) is optically stable at the tin atom for

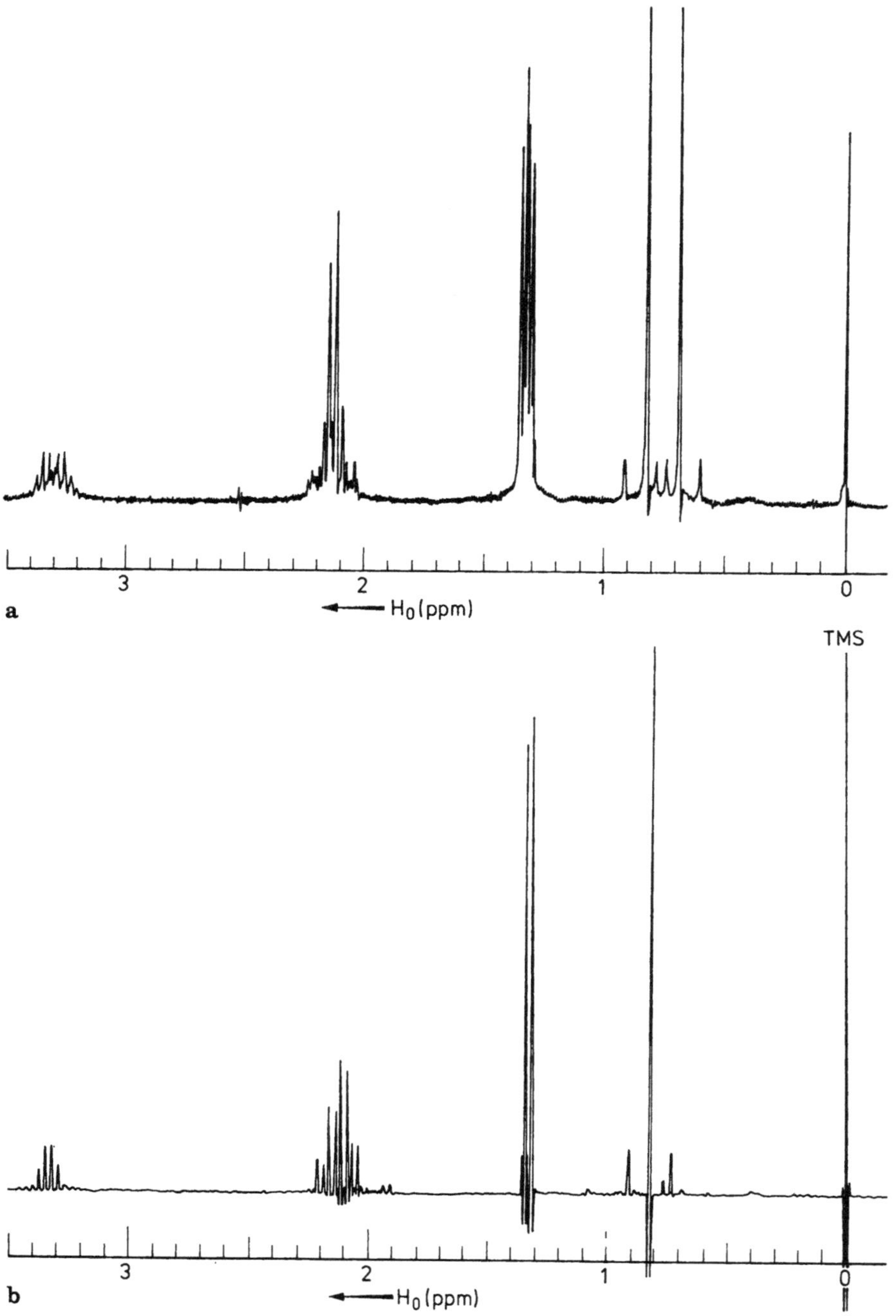

Fig. 6. 270 MHz ^{1}HMR spectra of the aliphatic part of (**a**) 50/50; (**b**) > 99/1 mixtures of methyl-phenyl(2-phenylpropyl)stannyltriphenylphosphinetricarbonylcobalt $(76)_A + (76)_B$ [18]. (taken from Ref. [18] with permission)

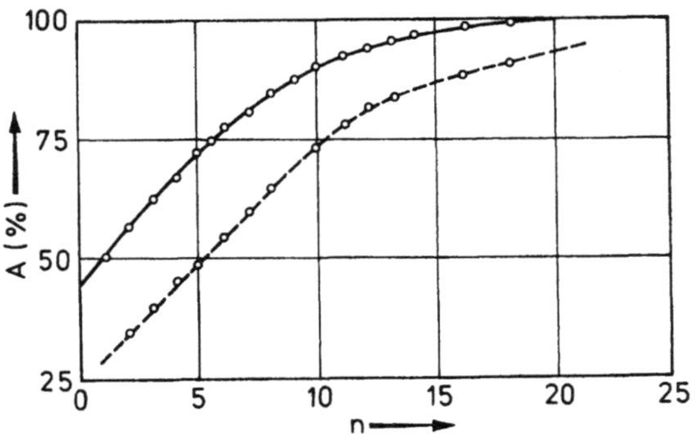

Fig. 7. Evolution of the diastereomeric composition of methylphenyl-(2-phenylpropyl)stannyltriphenylphosphinetricarbonylcobalt (*76*) in function of the number *n* of fractional recrystallizations in n-hexane [18]. ——— composition of the crystals; – – – – composition of the mother liquor. (taken from Ref. [18] with permission)

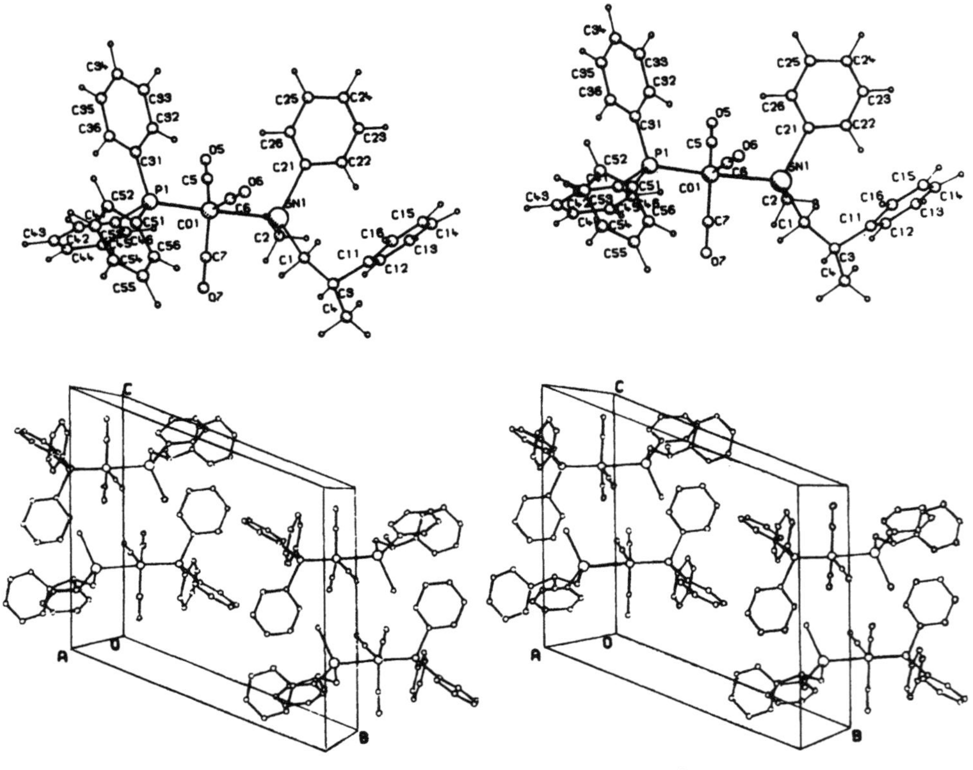

Fig. 8. Spectroscopic view and crystal packing of compound $(76)_A$ [57]. (taken from Ref. [57] with permission)

very long periods even in the presence of pyridine, whereas this was not the case for the cobalt complex (*76*) discussed before.

For compound (*78*), the 45:55 mixture obtained could be transformed by column chromatography [18] into two fractions containing respectively 40:60 and 50:50 of $(78)_A$:$(78)_B$. Here again, the A/B ratios remain unchanged for long periods in nonnucleophilic solvents but the addition of small quantities of pyridine or

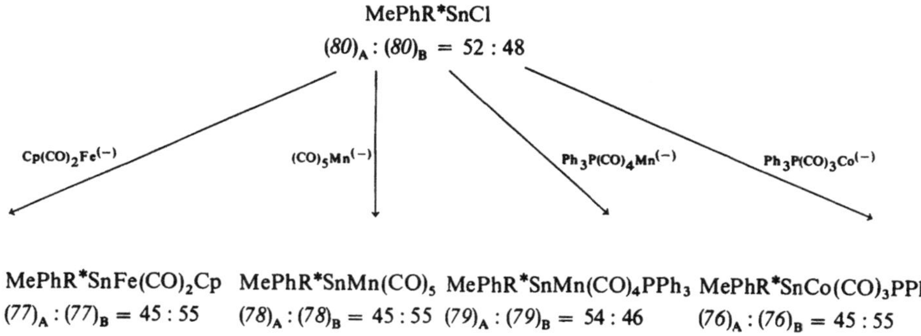

Fig. 9. Synthesis of methylphenyl(2-phenyl)stannyl-transition metal complexes [18].
($R^* = PhMeCHCH_2$)

DMSO causes an epimerization by which both fractions are transformed in a few hours into an equilibrium mixture (48:52).

For compound (*79*), the addition of pyridine or of DMSO does not cause any epimerization after a few hours. However, after three days, the $(79)_A:(79)_B = 54:46$ mixture obtained via scheme 9 is tranformed into the 52:48 equilibrium mixture.

The optical stability of methylphenyl(2-phenylpropyl)stannyl-transition metal complexes decreases as follows:

$$Cp(CO)_2Fe > Ph_3P(CO)_4Mn > Ph_3P(CO)_3Co > (CO)_5Mn$$ [18]

Methyl-1-naphthylphenylstannyl(diphenyl-N-methyl-N-(S)-1-phenylethylamino-phosphine)tetracarbonyl manganese (*81*)

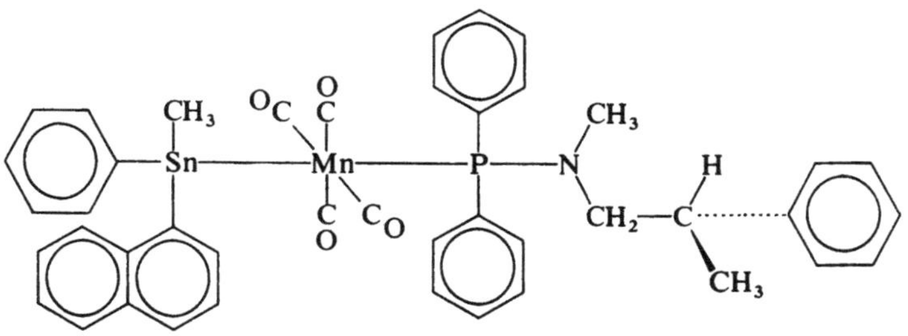

(17)

(*81*)

has been prepared and separated by fractional recrystallization into two fractions with $[\alpha]_{546}^{30^\circ} = +40$ and -71.4 respectively [58]. Unfortunately, these have identical NMR and IR spectra so that no information is available about their compositions [58]. However, the fact that the [α] of these fractions remain unchanged shows that (*81*) is configurationally stable at tin for long periods.

Methylphenyltritylstannyl(diphenyl-N-methyl-N-(S)-1-phenylethylaminophos-phine)tricarbonylcobalt (*82*)

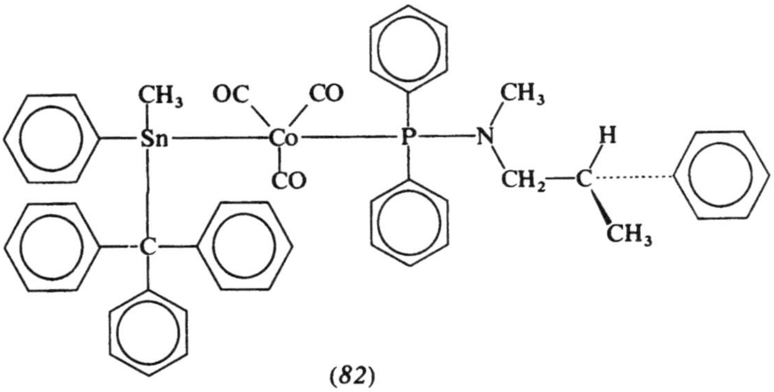

(82) (18)

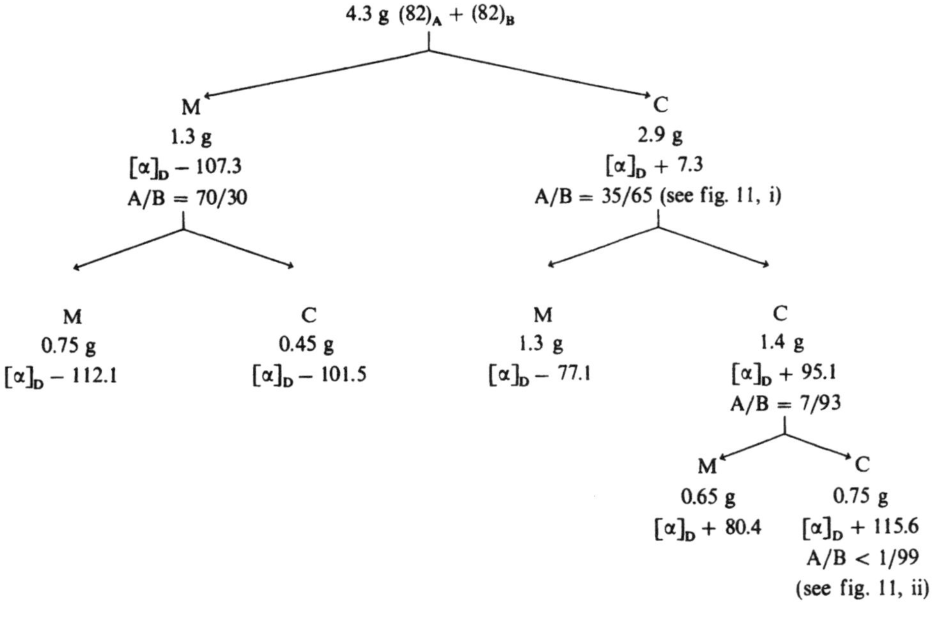

Fig. 10. Recrystallizations of $PhMe(Ph_3C)SnCo(CO)_3[(S)\text{-}Ph_2PNMeCHMePh]$, (*82*), in n-hexane [57]; M = evaporated mother liquor; C = crystals

has been prepared analogously [57]. The less soluble of the diastereomers of (*82*), $(82)_B$, was obtained in pure form after three recrystallizations from n-hexane (see Fig. 10). Its NMR spectrum is given in Fig. 11, ii). An X-ray diffraction study has shown [57] that it is the **threo**-mixture (see Fig. 12). This is an example of the case where the asymmetric carbon atom is resolved and where the two diastereomers could be separated.

As the absolute configuration of the asymmetric carbon atom is known to be S, it can be concluded from the relative configuration [56] of both chiral centres that $(82)_B$ is the S,S-isomer. As far as we know, this is the first absolute configuration

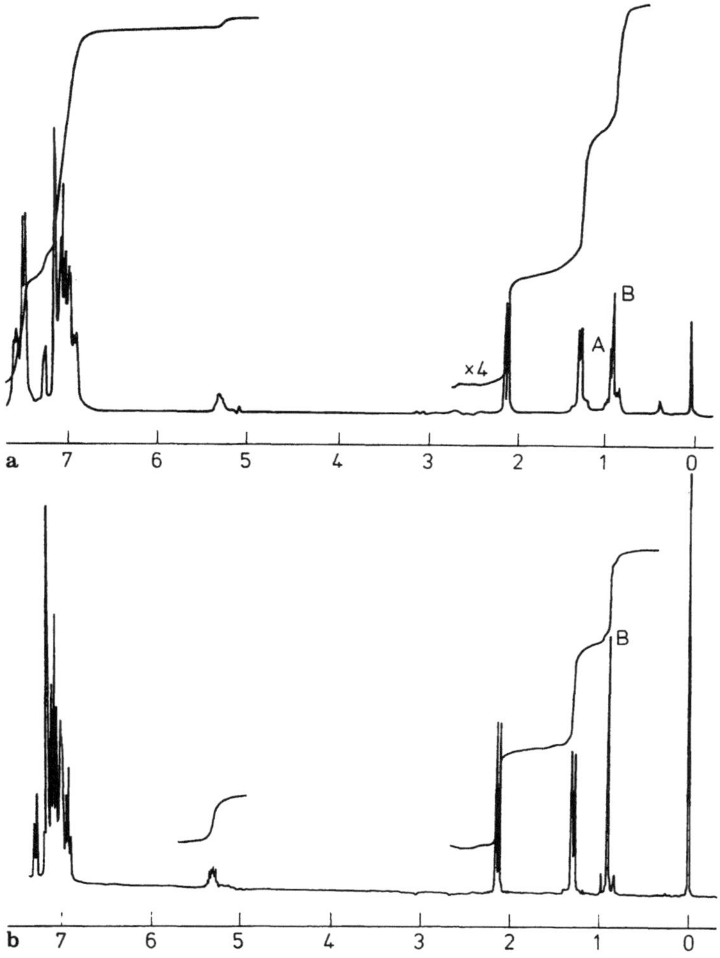

Fig. 11. 270 MHz ^{1}HMR spectra of PhMe(Ph$_3$C)Sn-Co(CO)$_3$[(S)-Ph$_2$PNMePh], (*82*). **a)** A/B = 35/65; **b)** A/B < 1/99 [57] (see text). (taken from Ref. [57] with permission)

determination around a four-coordinated tin atom [57]. The helicoidal conformation of the trityl group of $(82)_B$ observed in the solid state but not in solution [89] has also been observed for methylphenyltrityltin bromide (*83*) [90,91] (see Fig. 13).

4.2 The Use of Optically Active Compounds

Section 3 shows that many optically active organotin compounds with an asymmetric tin atom as only chiral center can be made. This fact is already strong evidence for the optical stability of those compounds. Furthermore, their optical rotation does

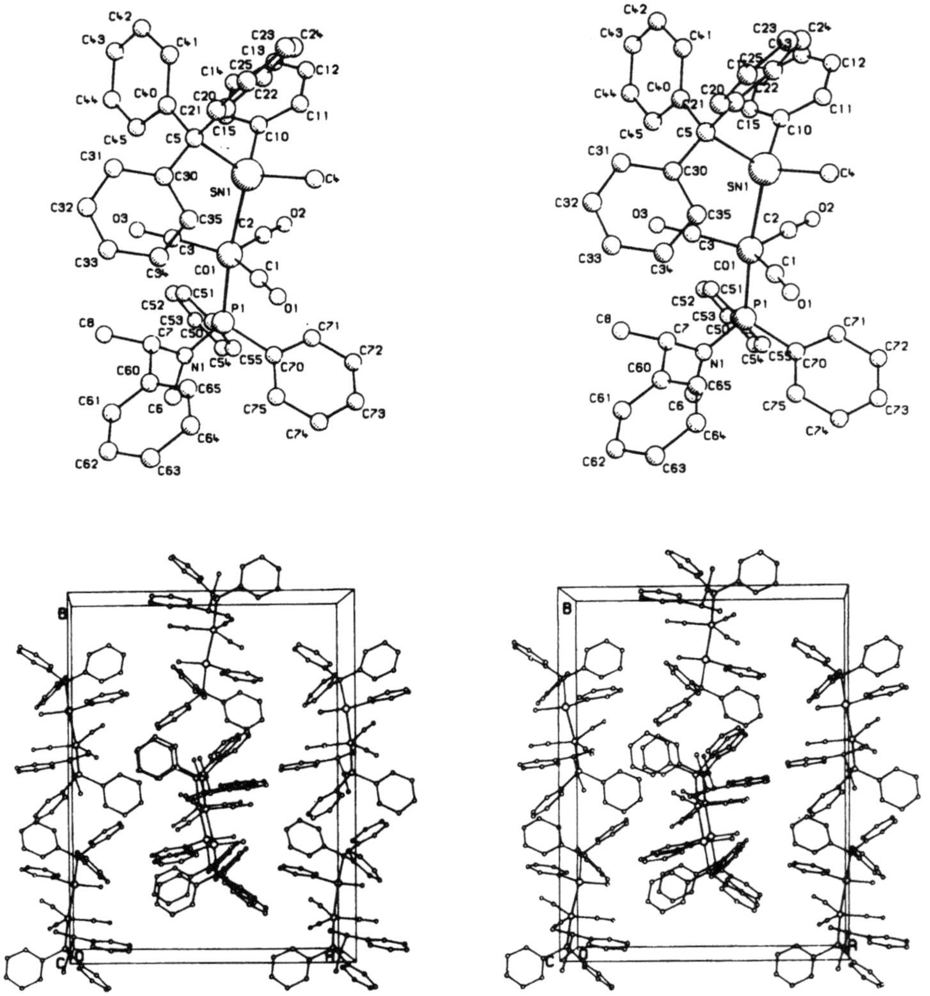

Fig. 12. Stereoscopic view and crystal packing of compound $(82)_B$ [57]. (taken from Ref. [57] with permission)

often not change even after long periods either in the presence or in the absence of nucleophilic species. If a change in [α] is detected, then a quantitative evaluation of their optical stability can be obtained.

4.2.1 Optical Stability of Triorganotin Hydrides

Methyl-1-naphthylphenyltin hydride (*66*) is an optically stable compound: the crystals can be kept in the dark at room temperature for 4 months or at 50 °C

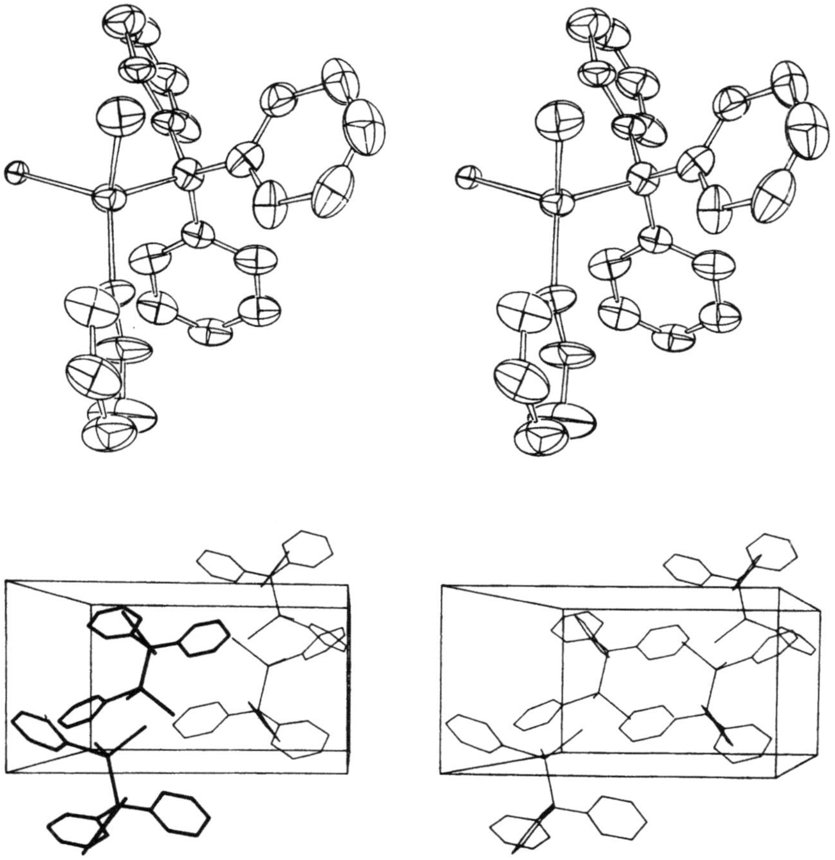

Fig. 13. Stereoscopic view of methylphenyltrityltin bromide (*83*)[90,91], and of the crystal packing of compound (*83*)[91]

for 10 minutes and even at 60 °C (above its melting point) for 8 minutes without any measurable variation of its optical rotation[44].

The optical rotation of methylneophylphenyltin hydride (*12*) remains unchanged if it is kept neat at −30 °C for 14 months or in benzene at 80 °C for 2 hours in the presence of hydroquinone, whereas a benzenic solution of (*12*) is fully racemized after 30 minutes at 80 °C in the presence of AIBN; 50% racemization is observed when the same solution without AIBN is kept for 17 days at room temperature. The optical rotation of a benzenic solution of (*12*) remains unchanged for weeks in the presence of hydroquinone at room temperature; (*12*) racemizes very slowly in ether or in benzene in the absence of hydroquinone and somewhat faster in benzene in the presence of AIBN at room temperature[44].

A radical mechanism is quite resonable to explain the optical instability of triorganotin hydrides [44], via the inversion [reaction (b)] of the triorganostannyl radical which is known to be non-planar [59].

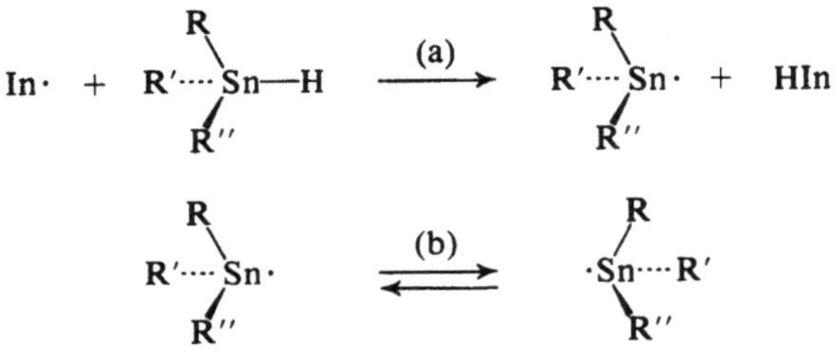

(19)

In the presence of nucleophilic solvents, a racemization is observed too: in less than one hour, in the presence or in the absence of hydroquinone, compound (*12*) is fully racemized in methanol [44]. A mechanism analogous to that proposed to rationalize the optical instability of triorganotin halides can be given here [44].

5 Stereoselectivity of Substitution Reactions at Tin

5.1 From Triorganotin Hydrides to Tetraorganotins

5.1.1 Stereoselective Insertion of a Methylene Group Between Tin and Hydrogen

The first example of a stereoselective substitution at tin was the reaction of (—)-*t*-butylneophylphenyltin hydride (*65*) ($[\alpha]^{30}_{365}$ — 0.9) with diazomethane in the presence of copper in diethyl ether to form optically active methylneophylphenyl-*t*-butyltin (*84*) ([α^{30}_{365} — 1.5) [20, 44, 60].

It is interesting to compare this case to the reaction of ethyl diazoacetate with chiral methyl-1-naphthylphenylsilane, which proceeds with at least 95% retention of configuration [61].

5.1.2 Hydrostannation of Olefinic Compounds

The hydrostannation of olefinic compounds in the presence of azobisisobutyronitrile AIBN is quite a well known reaction in organotin chemistry. Its mechanism is given below

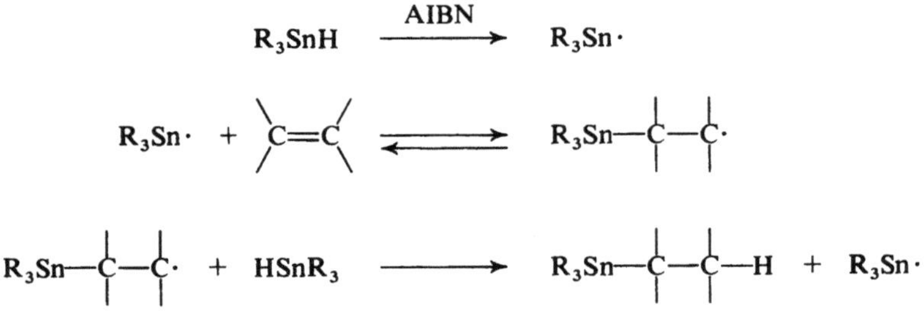

(20)

The reaction of (+)-methylneophylphenyltin hydride (*12*) with allyl alcohol takes 3 h. at 100 °C. Knowing that (*12*) is fully racemized after 30 minutes at 80 °C in the presence of AIBN, we are not surprised to notice that the adduct (*85*) obtained is not optically active [44].

$$\begin{array}{c} \quad Me \\ \quad | \\ PhMe_2CCH_2{-}Sn{-}CH_2CH_2CH_2OH \\ \quad | \\ \quad Ph \end{array}$$

(*85*)

Therefore, another analogous reaction was studied with a more reactive olefin, *viz.* methyl acrylate, which reacts with (+)-methylneophylphenyltin deuteride (*86*) **at room temperature** and yields after 18 h again an optically inactive adduct which is reduced with lithium aluminum hydride to give racemic isotopically labeled (*85*) [44]. After 18 h in the presence of AIBN at room temperature, (*86*) only loses 30% of its optical activity in benzene. The fact that the obtained adduct is optically inactive might be due to the nucleophilicity of methyl acrylate, which might be important enough to cause the racemization of (*86*).

An even more reactive olefin, bearing no heteroatoms which might be nucleophilic enough to cause the racemization of the triorganotin hydride, is bifluorenylidene. It reacts with (+)-(*12*) ($[\alpha]_D^{30}$ + 1.8) in the presence of AIBN to give (+)-bifluorenylylmethylneophylphenyltin, (+)-(*87*) ($[\alpha]_D^{30}$ + 0.8) [44,60]. Analogously, (—)-(*12*) ($[\alpha]_{365}^{30}$ — 0.7) is transformed into (—)-(*85*) ($[\alpha]_{365}^{30}$ — 0.4). The structure of (*87*), as determined by X-ray analysis is given in Fig. 14 and 15 [90].

The fact that this reaction is stereoselective implies that the chiral triorganostannyl radical formed in the first step is trapped by the olefin more rapidly than it inverts. This shows the optical stability of triorganostannyl radicals, which are known to be non-planar [59] like triorganogermyl radicals [62].

Analogous stereoselective addition reactions of triorganogermanes to double bonds have been described [63,64]: they proceed with retention of configuration.

5.2 From Triorganotin Hydrides to Hexaorganoditins

(+)-(*12*) ($[\alpha]_{365}^{30}$ + 13.2) is exothermically transformed into the corresponding optically active hexaorganoditin compound (—)-(*72*) ($[\alpha]_{365}^{30}$ — 28.9) [65]. This is the first example of a chiral hexaorganoditin, which is optically stable for weeks at room temperature (and which is in fact a mixture of a meso compound and of a non-racemic dl mixture, as seen by NMR). The same reaction with a mixture of (*12*) and triorganogermanium hydrides yielded only (*72*).

The reaction of (—)-(*65*) ($[\alpha]_{365}^{30}$ — 0.95) with dimethylmercury, which does not occur in the dark, yields, in daylight, the corresponding optically inactive mixture of meso and racemic dl hexaorganoditins (*88*) [44].

$$\underset{(-)\text{-}(65)}{(PhMe_2CCH_2)(Me_3C)PhSnH} \xrightarrow[h\nu]{Me_2Hg} \underset{(88)\text{, optically inactive}}{[(PhMe_2CCH_2)(Me_3C)PhSn]_2}$$

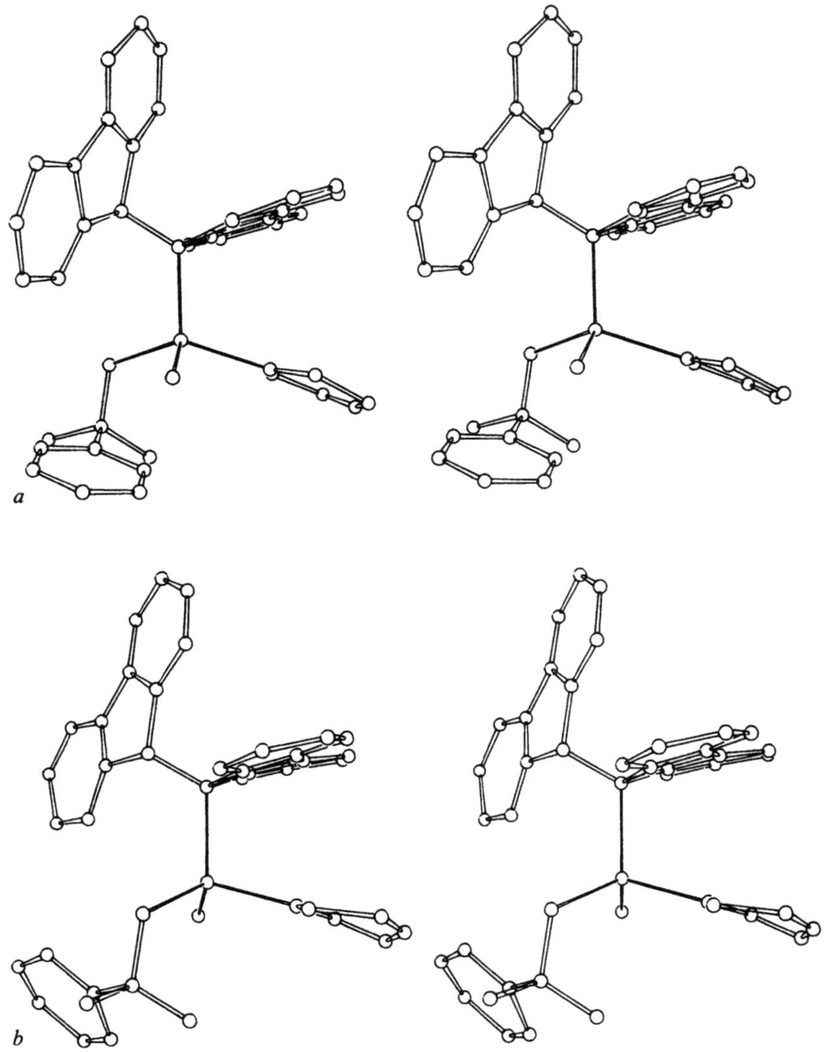

Fig. 14a and b. Stereoscopic view of the two conformations present in the trilinic crystals of (*87*)[93]

5.3 From Tetraorganotins to Hexaorganoditins

An optically inactive mixture of meso and racemic dl hexaorganoditins (*89*) is also obtained when (+)-methylneophyl-i-propyltrityltin, (+)-(*63*) ($[\alpha]_D^{30}$ + 1.8) reacts with lithium aluminum hydride[44]:

$$(PhMe_2CCH_2)(Me_2CH)MeSnCPh_3 \xrightarrow{LiAlH_4} [(PhMe_2CCH_2)(Me_2CH)MeSn]_2$$

(+)-(*63*) (*89*), optically inactive

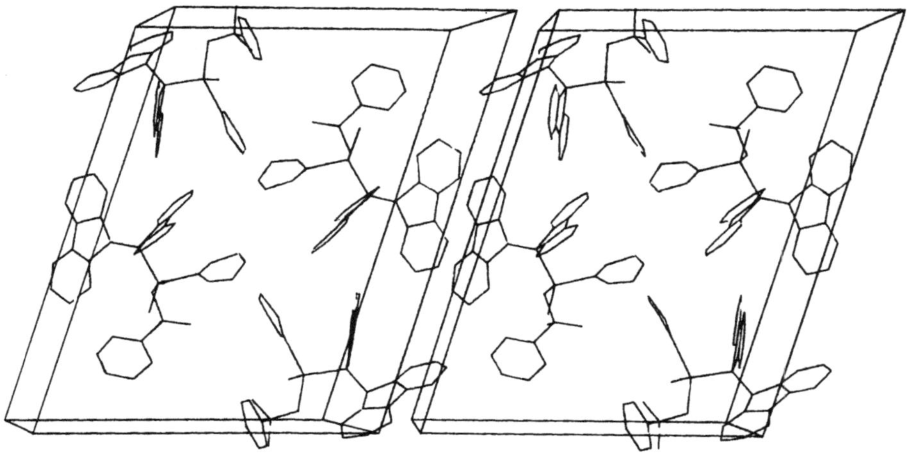

Fig. 15. Spectroscopic view of the crystal packing of compound (*87*) [93]

An analogous reaction has been observed between (+)-methylneophylphenyltrityltin, (+)-(*67*) and triethylhydroborate, giving optically inactive $[Me(PhMe_2CCH_2)\text{-}PhSn]_2$ (*72*) in 63% yield [66].

The trityl-tin bond is cleaved by triphenylstannyllithium: (+)-(*67*) reacts with Ph_3SnLi to give 54% Ph_6Sn_2, 50% optically inactive (*72*) and 30% racemic $Me(PhMe_2CCH_2)PhSnSnPh_3$ (*73*) [66].

5.4 From Triorganostannyl-Transition Metal Complexes to Hexaorganoditins

Analogous results have been obtained for the reaction between (methylneophylphenylstannyl)tricarbonyl(triphenylphosphinecobalt) (*15*) and Ph_3SnLi [67]: a mixture of 38% Ph_6Sn_2, 46% (*72*) and 45% (*73*) is obtained together with $[Co(CO)_3PPh_3]_2$. $HBEt_3^{(-)}$ anions do cleave the tin-cobalt bond too and the corresponding hexaorganoditin is formed here again [67].

The one-electron transfer mechanism described below can explain why comparable amounts of the three possible hexaorganoditins are formed when Ph_3SnLi reacts with RR'R"SnY substrates (Y = $Co(CO)_3PPh_3$, CPh_3 or Cl) [66,67].

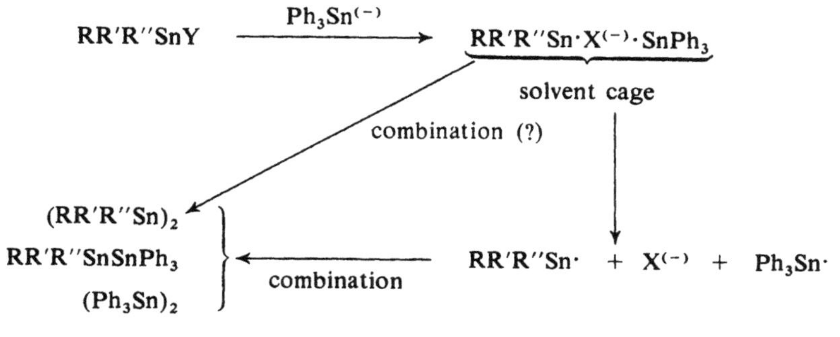

(21)

The obtention of optically inactive hexaorganoditin compounds from optically stable chiral substrates can be rationalized too, assuming that the formed triorganostannyl radical can invert if it is not trapped rapidly enough.

The reactions with $HBEt_3^{(-)}$ can be rationalized analogously if RR′R″SnY (Y = $Co(CO)_3PPh_3$, CPh_3) is transformed into RR′R″$Sn^{(-)}$ and if the formed RR′R″$Sn^{(-)}$ reacts further with RR′R″SnY according to the above scheme [66, 67].

5.5 From Triorganotin Hydrides to Triorganotin Halides

The reaction between triorganotin hydrides and organic halides is known since 1957 [68]. Analogous reactions have been described for chiral triorganosilicon [69] and -germanium [70] hydrides, which are stereoselective. It is well known that they proceed *via* the triorganometal radical, which can also be formed photochemically from metallated ketones [71].

A possible mechanism proposed by Kuivila was based on the fact that retardation by hydroquinone has been observed [72] (see however [73]) and that optically active alkyl halides RX have been transformed into racemic RD [72].

$$\left\{\begin{array}{l} R_3SnH + Q\cdot \longrightarrow R_3Sn\cdot + QH \\ RX + R_3Sn\cdot \longrightarrow [R\cdots X\cdots SnR_3]^{\ddagger} \longrightarrow R\cdot + XSnR_3 \\ R\cdot + HSnR_3 \longrightarrow RH + R_3Sn\cdot \end{array}\right. \quad (22)$$

Blackburn and Tanner have more recently proposed another mechanism [74] going through a triorganostannyl cation which must be planar.

$$\left\{\begin{array}{l} RX + R_3Sn\cdot \rightleftharpoons RX\cdot^{(-)} + R_3Sn^{(+)} \\ RX\cdot^{(-)} \longrightarrow \cdot R\cdot + X^{(-)} \end{array}\right. \quad (23)$$

The radical character of the reaction between (+)-(*12*) ($[\alpha]_{365}^{30}$ + 14) and CCl_4 is shown by its inhibition by hydroquinone [75]. Furthermore, the initial length of the induction period observed by IR, NMR or ORD [75] is an increasing function of the amount of air dissolved in the triorganotin hydride. Induction periods ranging from 3 to 300 minutes have been obtained according to the time of exposure to air of a sample of (*12*), initially degassed at 0.1 Torr for 45 minutes [75]. Nevertheless, the hydrostannolysis of CCl_4

$$\underset{(12)}{Me(PhMe_2CCH_2)PhSnH} + CCl_4 \rightarrow \underset{(3)}{Me(PhMe_2CCH_2)PhSnCl} + HCCl_3$$

is stereoselective: it gives an optically active triorganotin chloride (*3*) which is optically unstable: the lifetime of optically active 0.18 M (*3*) in CCl_4 is about 10 minutes [75]. Here again, the chiral triorganostannyl radical is trapped much more rapidly by CCl_4 than it inverts.

A trapping experiment of the optically active triorganotin chloride (*3*) with isopropylmagnesium bromide did give the expected methylneophylphenylisopropyltin (*9*) in a 20% yield but unfortunately, this tetraorganotin compound, which is optically stable (see Table 1), was obtained as a racemic mixture.

It might be interesting to comment on the difference between the configurational stability of (*12*) determined by NMR ($\tau = 0.045$ seconds $[(12)] = 2$ M [8,10]) and the optical stability of (*12*) determined by ORD ($\tau = 600$ seconds, for $[(12)] = 0.18$ M [75]). The very strong dependence of the racemization rate upon the concentration of (*12*) in this concentration range can be explained if this racemization is an overall **fifth** order process [75]. This very high reaction order is rather peculiar (cf. 2.2.) but not unexpected: using NMR line-shape analysis of a mixture benzylmethylneophyltin chloride (*90*) and iodide (*91*) in CCl_4, at concentrations of about 0.5 and 0.4 M respectively, Stynes and Allred [76] have shown that the inversion of configuration at tin occurs at the same rate as the halogen-halogen exchange and postulated the same associative mechanism for both processes. The racemization of (*90*) is an overall second-order process between 0.5 and 1.0 M (see also ref. [77]) but a third-order process has to be postulated between 1.0 and 1.6 M. Redl [78] suggested the formation of trimers and even of higher aggregates of R_3SnX to explain the Br—I exchange between trimethyltin bromide and iodide in cyclohexane. He proposes a rate law $v = k\,[Me_3SnI]^{\alpha}[Me_3SnBr]^{\beta}$ with values for α and β as high as 1.9 and 2.6 respectively.

5.6 From Tetraorganotins to Triorganotin Halides

The protodemetallation of optically active methylneophylphenyltrityltin, (—)-(*67*), in diethylether yields optically inactive methylneophyltrityltin chloride (*5*) [79]. However, (*5*) is configurationally stable even in the presence of DMSO in pyridine (cf. Table 1) and methylneophylphenyltin chloride (*12*), which is already configurationally unstable in the presence of traces of pyridine (cf. Table 1) racemizes in about 10 minutes in CCl_4 in the absence of pyridine

$$Me(PhMe_2CCH_2)PhSnCPh_3 \xrightarrow[Et_2O]{HCl/} Me(PhMe_2CCH_2)(Ph_3C)SnCl$$

(—)-(*67*), $[\alpha]_D^{30}$-40 (±)-(*5*)

Either the reaction is not stereoselective and yields a racemic mixture, or the racemization of the formed optically active (*5*) is very rapid in the presence of HCl [79] (cf. Fig. 2)

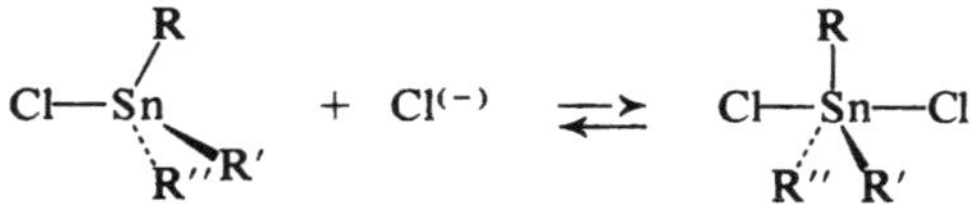
(24)

achiral

5.7 From Triorganotin Hydrides to Triorganostannyl-Transition Metal Complexes

Triorganotin hydrides do react with $HMn(CO)_5$ to give $Mn_2(CO)_{10}$ and the expected (triorganostannyl)pentacarbonyl manganese [80]. In THF, only these two compounds are formed. In benzene, in addition to these two products, the hexa-organoditin and some starting product are found in the reaction mixture.

Unfortunately, the reaction does not give any optically active (triorganostannyl)-pentacarbonyl manganese when performed on an optically active triorganotin hydride. In boiling THF, $HMo(CO)_5$ and (+)-(*12*) $[(\alpha]^{30}_{546}$ + 2) are converted after 2 hours into optically inactive (*14*) even if (*14*) is optically stable for long periods in the absence of very strong nucleophiles like DMSO (see Table 1). In boiling benzene, the same reagents are converted into optically inactive (*14*) (30%), partially racemized (*12*) (25%) ($[\alpha]^{30}_{546}$ + 0.2) and optically inactive (*72*) (5%) [80].

5.8 From Tetraorganotins to Triorganostannyl-Transition Metal Complexes

The trityl-tin bond is cleaved by $Cp(CO)_2Fe^{(-)}$: (−)-(*67*) is then converted into racemic (methylneophylphenylstannyl)dicarbonylcyclopentadienyl iron, (±)-(*13*) in 83% yield [66].

$$(-)\text{-Me(PhMe}_2\text{CCH}_2\text{)PhSnCPh}_3 \xrightarrow[2)\ H_2O]{1)\ NaFe(CO)_2Cp} \text{Me(PhMe}_2\text{CCH}_2\text{)PhSnFe(CO)}_2\text{Cp}$$

(−)-(*67*), $[\alpha]^{30}_{D}$-*30* (±)-(*13*)

The absence of stereoselectivity observed here has been explained by a one-electron transfer mechanism [66] (see 5.4)

$$\text{Me(PhMe}_2\text{CCH}_2\text{)PhSnCPh}_3 \xrightarrow{Cp(CO)_2Fe^{(-)}} \underbrace{\text{Me(PhMe}_2\text{CCH}_2\text{)PhSn}\cdot\text{Ph}_3\text{C}^{(-)}\cdot\text{Fe(CO)}_2\text{Cp}}_{\text{solvent cage}} \quad (25)$$

$$\xrightarrow{\text{combination}} \text{Me(PhMe}_2\text{CCH}_2\text{)PhSn–Fe(CO)}_2\text{Cp} + \text{Ph}_3\text{C}^{(-)}$$

5.9 From Triorganostannyl-Transition Metal Compounds to Other Triorganostannyl-Transition Metal Complexes

Racemic **threo** [57]-[methylphenyl(2-phenylpropyl)stannyl]tricarbonyl(triphenylphosphine) cobalt (*76*) reacts with $NaFe(CO)_2Cp$ to give the same 45/55 mixture of [methylphenyl(2-phenylpropyl)stannyl]dicarbonylcyclopentadienyl irons $(77)_A/(77)_B$

as the one obtained (see Fig. 9) from a 52/48 mixture of the diastereomeric optically unstable $(80)_A/(80)_B$ [18].

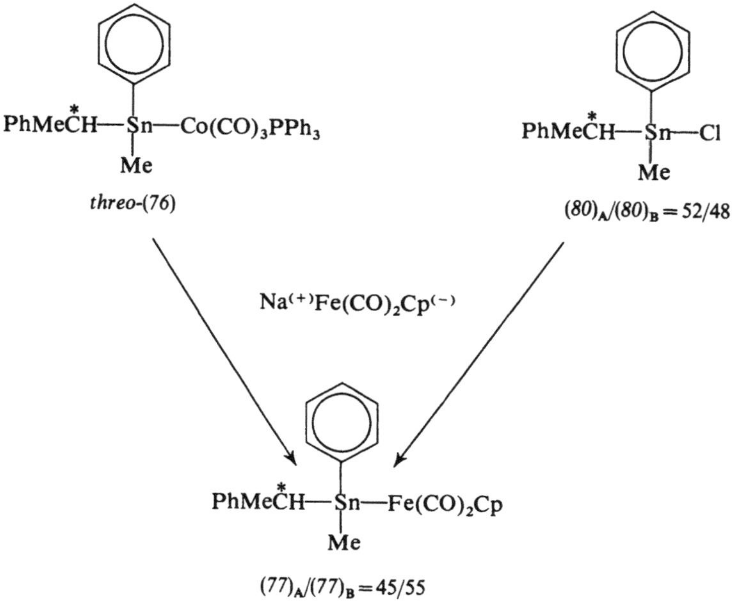

(26)

Epimerization at tin has occured during the substitution process. A one-electron transfer mechanism has been proposed here as for the reactions described in Sections 5.3, 5.4, and 5.8 to account for the absence of stereoselectivity.

5.10 Hydrogen-Deuterium Exchange

(+)-(*12*) reacts with lithium aluminum deuteride in diethyl ether at room temperature. After 5 hours, the optical activity is reduced by 61% and 47% of (*12*) is converted into the corresponding deuteride (*86*) [44].

Analogously, the ratio % exchange: % inversion for the reaction between (+)-(*86*) and $LiAlH_4$ is 2.2 after 1 h, 3.0 after 2 h, 3.2 after 5 h. and 1.4 after 24 h [44]. Similar results were obtained by Parker [81] for the analogous reaction between (+)-methyl-1-naphthylphenylsilane and $LiAlD_4$ in diglyme/dioxane (1/1); he obtained ratios of 2.4 after 34 min., 2.7 after 65 min., 2.5 after 90 min. and 1.7 after 150 min. This suggests that two parallel reactions are operative, one proceeding with retention of configuration and the other with racemization. Lequan [82] came to analogous conclusions for the reaction between (+)-methyl-1-naphthylphenylsilane and $LiAlD_4$ in THF. In contrast, this last H—D exchange proceeds with retention of configuration in diethyl ether [83].

This might explain why one equivalent of the chiral reducing agent of Vigneron and Jacquet [45] (cf. 3.3.) transforms compound (*3*) into optically active (*12*) ($[\alpha]^{30}_{365}$ + 7.4) with 80% yield whereas, with 8 equivalents of this reducing agent, the chemical yield is increased up to 93% but (*12*) is obtained as a racemic mixture.

The H—D exchange between triethyltin deuteride and di-isobutylaluminum hydride has been studied by Neumann [84]. A four-center transition state has been proposed for this reaction

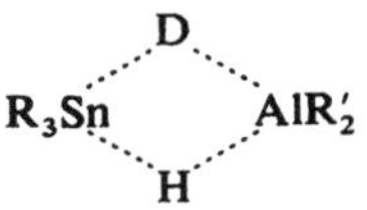

(27)

However, when (+)-methylneophylphenyltin deuteride, (+)-(*86*) ($[\alpha]^{30}_{365}$ + 10.7) is kept in the dark mixed with five equivalents of diethylaluminum hydride for ten hours at room temperature in benzene, optically inactive (*12*) is formed [44]. (In the absence of $(Et_2AlH)_2$ less than 3% of (*12*) is racemized under these conditions). The four-center transition state is therefore very unlikely.

Neumann has described the hydrogen-deuterium exchange between triisobutyltin hydride and triethyltin deuteride [24]. Here again, a four-center mechanism has been postulated. (+)-Methylneophylphenyltin deuteride, (+)-(*86*) ($[\alpha]^{30}_{365} = +5.8$), which loses about 7% of its optical activity when heated at 40 °C for 9 hours, is converted after 9 hours at 40 °C in the presence of 5 equivalents of triphenyltin hydride, into a mixture containing 76% of the corresponding hydride (*12*) and 24% of (*86*), characterized by $[\alpha]^{30}_{365} = +5$, showing that this H/D exchange proceeds with at least 93% retention (at most 7% racemization) [44]. A four-center transition state is fully consistent with the high stereoselectivity observed in this case. However, hexaphenylditin has been isolated from the reaction mixture [44]. Analogous but catalyzed H—D exchange between two triorganosilanes [85] or between two triorganogermanes [86] proceed with retention of configuration at the metal atom too.

6 Conclusion

Many organotin compounds are optically stable for long periods so that all the reactions transforming one of them into another can be studied from the stereochemical point of view. Several stereoselective reactions at tin have already been found. However the stereochemistries and even the stereoselectivities have not yet been determined (if the H—D exchange is not taken into account) because the optical purities (and a fortiori the absolute configurations) of the starting materials and of the reaction products were not known. The recent development of inclusion chromatography on microcrystalline cellulose triacetate has allowed the separation of the enantiomers of racemic mixtures of organotin compounds and determination of the optical purity of such derivatives. Therefore, the study of the stereoselectivity and even of the stereochemistry, of substitution reactions at the tin atom has become a possible goal for further studies in this field.

7 Acknowledgement

I have avoided the problem of referring to my research students by name by mentioning none of them; but their experimental work is gratefully acknowledged, and their names can be found in the references given below. I am indebted to them for their indispensable and stimulating collaboration.

8 List of Symbols and Abbreviations

acacH	acetylacetone
AIBN	azobisisobutyronitrile
Cp	η^5-cyclopentadienyl
DMSO	dimethylsulfoxide
$\Delta\nu_{(AB)}$	chemical shift difference (between A and B) expressed in Herz
$\Delta\sigma$	chemical shift difference expressed in ppm
Et	ethyl
HMPA	hexamethylphosphotriamide (hexametapol)
INV	inversion
IR	infra-red spectroscopy
Me	methyl
NMR	nuclear magnetic resonance spectroscopy
ORD	optical rotatory dispersion
Ph	phenyl
3,5-Me_2Ph	3,5-dimethylphenyl
THF	tetrahydrofurane
τ	lifetime

9 References

1. Sommer, L. H., Frye, C. L.: J. Amer. Chem. Soc. *81*, 1013 (1959)
2. Schwarz, R., Lewinsohn, M.: Chem. Ber. *64*, 2352 (1931)
3. a) Bott, R. W., Eaborn, C., Varma, I. D.: Chem. Ind., *1963*, 614;
   b) Eaborn, C., Simpson, P., Varma, I. D.: J. Chem. Soc. A, *1966*, 1133
4. Brook, A. G., Peddle, G. J. D.: J. Amer. Chem. Soc., *85*, 1869, 2338 (1963)
5. a) Pope, J., Peachey, S. J.: Proc. Chem. Soc. (London) *16*, 42, 116 (1900)
   b) Weyer, J.: Angew. Chem., Internat. Edit. *13*, 596 (1974)
6. Kipping, F. B.: J. Chem. Soc. *131*, 2365 (1928)
7. Naumov, C. N., Manulkin, Z. M.: Zh. Obshch. Khim. *5*, 281 (1935); Chem. Abstr. *29*, 5071 (1935)
8. Peddle, G. J. D., Redl., G.: Chem. Comm. 1968, 626
9. Belloli, R.: J. Chem. Educ. *46*, 640 (1969)
10. Peddle, G. J. D., Redl, G.: J. Amer. Chem. Soc. *92*, 365 (1970); Redl, G.: J. Organometal. Chem. *22*, 139 (1970)
11. Folli, U., Iarossi, D., Taddei, F.: J. Chem. Soc. Perkin II (1973), 638, 1284
12. Gielen, M., Mokhtar-Jamaï, H.: J. Organometal. Chem. *91*, C33 (1975)
13. Gielen, M., Mokhtar-Jamaï, H.: Bull. Soc. Chim. Belg. *84*, 1037 (1975)
14. Gielen, M. et al.: Recl. Trav. Chim. Pays-Bas *88*, 1337 (1969)
15. Gielen, M.: Topics in Stereochemistry *12*, Chap. 5, 221, J. Wiley (1980)
16. Gielen, M. et al.: Bull. Soc. Chim. Belg. *83*, 337 (1974)
17. Gielen, M., Hoogzand, C., Vanden Eynde, I.: ibid. *84*, 939 (1975)
18. Gielen, M., Vanden Eynde, I.: Israel J. Chem. *20*, 93 (1980)
19. Gielen, M., Simon, S.: Bull. Soc. Chim. Belg. *86*, 589 (1977)
20. Gielen, M. et al.: Israel J. Chem. *15*, 74 (1977)
21. Gielen, M. et al.: J. Organometal. Chem. *34*, 315 (1972)
22. Gielen, M., Tondeur, Y.: Bull. Soc. Chim. Belg. *84*, 933 (1975)
23. Mokhtar-Jamaï, H.: Ph. D. Thesis, Free Univ. of Brussels U.L.B. (1975)
24. van Koten, G. et al.: J. Chem. Soc. Dalton 1980, 1352
25. van Koten, G., Noltes, J. G.: J. Amer. Chem. Soc. *98*, 5393 (1976)

26. van Koten, G., Jastrzebski, J. T. B. H., Noltes, J. G.: J. Organometal. Chem. *177*, 283 (1979)
27. van Koten, G. et al.: J. Amer. Chem. Soc. *100*, 5021 (1978)
28. Carré, F. H., Corriu, R. J. P., Thomassin, R. B.: J. Chem. Soc. D, (1968), 560
29. a) Corriu, R., Henner-Leard, M.: J. Organometal. Chem. *64*, 351 (1974); *65*, C39 (1974); *74*, 1 (1974);
b) Mc. Kinnie, B. G., Cartledge, F. K.: ibid. *104*, 407 (1976);
c) Cartledge, F. K., Mc. Kinnie, B. G., Wolcott, J. M.: ibid. *118*, 7 (1976)
30. Sommer, L. H., Stark, F. O., Michael, K. W.: J. Amer. Chem. Soc. *85*, 3898 (1963)
31. Gielen, M., Mokhtar-Jamaï, H.: J. Organometal. Chem. *129*, 325 (1977)
32. Corriu, R., Henner, M.: Bull. Soc. Chim. France (1974), 1447
33. Bos, K. D., Bulten, E. J., Noltes, J. G.: J. Organometal. Chem. *99*, 397 (1975)
34. Lequan, M., Meganem, F., Besace, Y.: ibid. *113*, C13 (1976)
35. see Ref. [1,3,4] and Gielen, M. et al., "The optical stability of organotin compounds", in "Organotin Compounds: new chemistry and applications", ed.: J. J. Zuckerman, Advances in Chemistry Series, *157*, 249 (1976)
36. Mokhtar-Jamaï, H. et al.: Proc. 5th Int. Conf. Organometal. Chem. (Moscow) *1*, 523 (1971); *2*, 359 (1971);
Mokhtar-Jamäï, H., Gielen, M.: Bull. Soc. Chim. France *9B* (1972), 32;
Gielen, M., Moktar-Jamäï, H.: Annals N.Y. Acad. Sci. *239*, 208 (1974); Bull. Soc. Chim. Belg., *84*, 197 (1975)
37. van der Kooi, H. O., Wolters, J.: Third Int. Conf. on the organometallic a coordination chem. of germanium, tin a. lead, Abstract A.29: Chiral organolead compounds: synthesis a. characterization, Dortmund, 21–25 July 1980
38. Lequan, M., Meganem, F.: J. Organometal. Chem. *94*, C1 (1975)
39. Gielen, M., Simon, S.: Bull. Soc. Chim. Belg. *86*, 39 (1977)
40. Lequan, M., Queroix, S.: J. Organometal. Chem. *194*, 33 (1980)
41. Gielen, M.: Annual Reports Inorg. Gen. Synth. (1976) (Ed. H. Zimmer) Academic Press 1977, p. 337
42. Lequan, R. M., Lequan, M.: J. Organometal. Chem. *202*, C99 (1980)
43. Gielen, M., Tondeur, Y.: ibid. *127*, C75 (1977)
44. Gielen, M., Tondeur, Y.: ibid. *169*, 265 (1979)
45. Vigneron, J. P., Jacquet, I.: Tetrahedron *32*, 939 (1976)
46. Gielen, M.: Pure a. Appl. Chem. *52* (3), 657 (1980)
47. McMaster, A. D., Stobart, S. R.: to be published
48. Stobart, S. R. et al.: 3rd Int. Conf. on the Organometallic a. Coordination Chem. of Ge, Sn and Pb, Abstract A32, Tetra(indenyl)tin: a stereochemically significant molecule, Dortmund, 21–25 July 1980
49. Gielen, M., Molenberghs, L., Vanden Eynde, I.: Bull. Soc. Chim. Belg. *90*, 177 (1981)
50. Mannschreck, A. et al.: Angew. Chem. Int. Ed. Engl. *19*, 469 (1980)
51. Hesse, G., Hagel, R.: Chromatographia *6*, 277 (1973); *9*, 62 (1976); Liebigs Ann. Chem. 996, (1976)
52. Gielen, M., Vanden Eynde, I.: Organometal. Chem. Libr. *12*, 193 (1981)
53. Vanden Eynde, I., Gielen, M.: J. Organometal. Chem. *198*, C55 (1980)
54. van der Kooi, H. O., Wolters, J., van der Gen, A.: Recl. Trav. Chim. Pays-Bas *98*, 353 (1979)
55. Vanden Eynde, I. et al.: Polyhedron, in press.
56. Gielen, M.: J. Chem. Educ. *54*, 673 (1977)
57. Gielen, M. et al.: Bull. Soc. Chim. Belg. *89*, 915 (1980)
58. Gielen, M., Vanden Eynde, I.: J. Organometal. Chem. *217*, 205 (1981)
59. Lehnig, M. et al.: Bull. Soc. Chim. Belg. *89*, 907 (1980);
see also Busschhaus, H. V., Lehnig, M., Neumann, W. P.: J. Chem. Soc. Chem. Comm. (1977), 129
and Watts, G. B., Ingold, K. V.: J. Amer. Chem. Soc. *94*, 491 (1972)
60. Gielen, M., Tondeur, Y.: J. Organometal. Chem. *128*, C25 (1977)
61. Brook, A. G., Duff, J. M., Anderson, D. G.: J. Amer. Chem. Soc. *92*, 7567 (1970)
62. Sakurai, H., Mochida, K.: J. Chem. Soc. Chem. Comm. (1971), 1581;
Sakurai, J. et al.: J. Organometal. Chem. *38*, 275 (1972)
63. Corriu, R. J. P., Moreau, J. J. E.: (1971), 821; J. Organometal. Chem. *40*, 55, 73 (1972)

64. Dubac, J. et al.: J. Organometal. Chem. *127*, C69 (1977)
65. Gielen, M., Tondeur, Y.: J. Chem. Soc. Chem. Comm. (1978), 81
66. Gielen, M., Vanden Eynde, I.: J. Organometal. Chem. *218*, 315 (1981)
67. Gielen, M., Vanden Eynde, I.: Transition Metal. Chem. *6*, 344 (1981)
68. van der Kerk, G. J. M., Noltes, J. G., Luijten, J. G. A.: J. Appl. Chem. *7*, 356 (1957);
see also Noltes, J. G., van der Kerk, G. J. M.: Chem. & Ind. (1959), 294;
Kupchik, F. J., Connolly, R. E.: J. Org. Chem. *26*, 4747 (1961);
Lorenz, D., Becker, E.: ibid. *27*, 3370 (1962);
Kriegsmann, H., Ullbright, K.: Z. Chem. *3*, 67 (1963)
69. Sakurai, H., Murakami, M., Kulada, M.: J. Amer. Chem. Soc. *91*, 519 (1969)
70. Sakurai, H., Mochida, K.: J. Chem. Soc. Chem. Comm. (1971), 1581
71. Brook, A. G., Duff, J. M.: J. Amer. Chem. Soc. *91*, 2118 (1969)
72. Kuivila, H. G., Menapace, L. W.: ibid. *86*, 3047 (1964)
73. Lorenz, D. H. et al.: J. Org. Chem. *28*, 2332 (1963)
74. Blackburn, E. V., Tanner, D. D.: J. Amer. Chem. Soc. *102*, 692 (1980)
75. Gielen, M., Tondeur, Y.: Nouv. J. Chim. *2*, 117 (1978)
76. Stynes, D. V., Allred, A. L.: J. Amer. Chem. Soc. *93*, 2666 (1971)
77. Ladd, J. A., Glasberg, B. R.: J. Chem. Soc. Dalton (1975), 2378 and the references cited therein
78. Redl, G., Winokur, M.: J. Organometal. Chem. *26*, C36 (1971)
79. Gielen, M., Vanden Eynde, I.: Bull. Soc. Chim. Belg. *90*, 195 (1981)
80. Gielen, M., Vanden Eynde, I.: Transition Met. Chem. *6*, 128 (1981)
81. Parker, G. A.: Ph. D. Thesis, Penn. State Univ. (1963)
82. Meganem, F., Jean, A., Lequan, M.: J. Organometal. Chem. *74*, 43 (1974)
83. Sommer, L. H.: "Stereochem., Mechanism and Silicon", McGraw Hill, N.Y. 1965, p. 102–104
84. Neumann, W. P., Sommer, R.: Angew. Chem., Intern. Ed. Engl. *2*, 547 (1963)
85. Sommer, L. H., Lyons, J. E., Fujimoto, H.: J. Amer. Chem. Soc. *91*, 705 (1969)
86. Corriu, R. J. P., Moreau, J. J. E.: J. Organometal. Chem. *40*, 55 (1972)
87. Gielen, M.: Accounts Chem. Res. *6*, 198 (1973)
88. Gielen, M., Vanden Eynde, I.: Bull. Soc. Chim. Belg. *90*, 101 (1981)
89. Vanden Eynde, I.: Ph. D. Thesis, Free Univ. of Brussels V.U.B. (1981)
90. Tondeur, Y.: Ph. D. Thesis, Free Univ. of Brussels U.L.B. (1979)
91. Germain, G. et al.: Bull. Soc. Chim. Belg., to be published
92. Lequan, R. M., Lequan, M.: Tetrahedron Lett. *22*, 1323 (1981)
93. Meunier-Piret, J., Van Meerssche, M.: Bull. Soc. Chim. Belg., to be published

Coordination Effects in Formation and Cross-Linking Reactions of Organotin Macromolecules

Zakir M. O. Rzaev

Academy of Sciences of the Azerbaijan SSR, Institute of Organochlorine Synthesis, Sumgait, USSR

Table of Contents

1 Introduction

In recent years considerable progress has been made in the synthesis of new organotin monomers and polymers and in their application in various branches of technology. The ever increasing interest in organotin compounds including organotin polymers may be attributed to their unique properties and the possibility to develop novel materials with specified characteristics.

The appreciable advances in the field of organotin compounds are associated with the elucidation of their mechanisms of formation, detection of new properties and their application as biocides, stabilizers, antifouling, and antimicrobic coatings, etc.

It has been known [1-5] that in many organotin compounds, including monomers and polymers, the tin atom is in a coordination state which is essential for the formation and cross-linking of organotin macromolecules and accounts for quite a number of specific properties of these polymers which are not inherent in organic isostructural analogs of other elements belonging to group IVB of the Periodical Table.

However, in many papers on polymerization and copolymerization of organotin monomers, the role of complex formation in elementary acts of polymer formation has been either ignored or considered inadequately.

Our research in this field is mainly carried out on: a) synthesis and conversion of oligoorganoepoxystannanes; b) free-radical copolymerization of organotin monomers with various vinyl monomers; and c) cross-linking of organotin macromolecules and development of protective polymeric coatings with specific properties on their basis.

A large number of studies on the polymerization and copolymerization of tin-containing monomers and the synthesis of new organotin polymers have been reported. In this review article the results are discussed and further ways for research in this interesting and important field of organometallic and polymer chemistry outlined.

Considering the fact that this review is the first attempt to generalize and systematize the data accumulated in this interesting and important area and admitting possible omissions, the author will be grateful to the reader for his constructive criticism.

2 Complex Formation

Organotin compounds including organotin monomers tend to form complexes with various electron-rich compounds and it can be assumed that pentacoordination is the basis for the chemistry of these materials. A tin atom has vacant d-orbitals the energy of which is close to that of s- and p-orbitals in the outermost electron shell. As a consequence, organotin compounds exhibit a variety of chemical and physical properties which differ from those of their organic analogs and can be attributed to d-orbitals in tin atom. Naturally, transition of a tin atom into an sp^3d or sp^3d^2 valence state and formation of coordination bonds occurs most readily in organotin compounds containing substituents with unshared electron pairs.

The tendency of organotin compounds to form complexes with various organic compounds bearing electron-rich groups has been corroborated experimentally [6-16].

Nasielsky [17] discussed both the static and dynamic stereochemistry of some organotin compounds and showed that pentacoordination is the dominant factor determining the physical and chemical properties of these compounds.

Complex formation in binary systems under the influence of organotin monomers (organotin carboxylate and epoxide monomers) was studied by IR and NMR techniques [2,3,18-23].

In the following are discussed some structural features of the initial organotin derivatives of unsaturated acids, which are characterized by the presence of coordination-bound associated molecules,

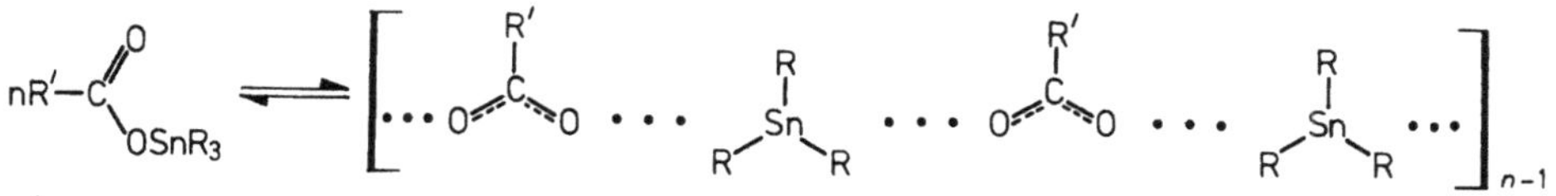

R′: $CH(CH_3)_2$, $CH{=}CH_2$, $C(CH_3){=}CH_2$, $C_6H_5CH{=}CH$

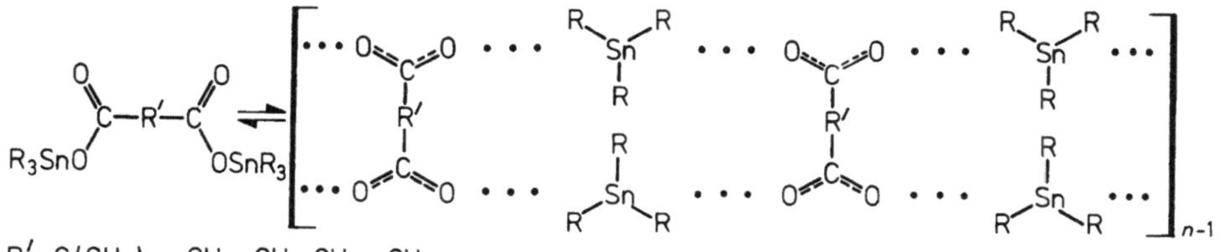

R′: $C(CH_3)_2$, $CH{=}CH$, $CH_2{-}CH_2$

R : alkyl, phenyl

It is seen from IR results on organotin methacrylates and their organic analog, methyl methacrylate (MMA), that longer alkyl substituents at the tin atom result on the one hand in more intensive absorption bands of the C=C bond, and on the other hand, in shifting of these bands in the R_3SnOOC group to a longer wavelength region (from 1590 to 1620 cm^{-1}). The latter is indicative of the effect of R_3Sn groups on the electronegativity of the carbonyl group conjugated with the double bond [2,19].

The spectrum of bis-triethylstannyl maleate (R′ = CH=CH) (bis-TESM) exhibits no absorption bands at 1640 and 1720 cm^{-1}, which is typical of the normal position of carbonyl groups, but shows an absorption band at 1566 cm^{-1} which corresponds to C=O group oscillation in the carboxylate ion. The spectrum of ô-phenyl-tri-butylstannyl acrylate (PBSA) displays bands qt 1642 and 1738 cm^{-1} (weak), which are characteristics of valency oscillation of a carbonyl group, and a band at 1576 cm^{-1} relating to carboxylate group (COO^-) oscillation. The appearance of the two different groups is probably due to the effect of the phenyl and butylstannyl groups on the double bond conjugation which, in turn, results in a variation of the polarity of the C=O group [24,92].

In the spectra of model compounds such as succinic anhydride (SA), tri-n-butyl-stannyl 2-methylpropanoate (TBMP) and their mixtures and a copolymer of tributylstannyl methacrylate (TBSM) with maleic anhydride (MA), a pronounced complexing is observed between the tin atom and the carbonyl group; the C=O absorption bands of the carboxylate ion disappear almost completely and the strength of the covalent C—Sn—O bond increases considerably due to coordinative interaction with carbonyl groups.

The IR data suggest an anomalous decrease of the D_{1575}/D_{1610} value with changing composition of the monomer mixture at constant total concentration of initial monomers. This concirms that the formation of a coordination complex between the carbonyl group and the tin atom at an equimolar TBSM-to-MA ratio is highly probable.

On the basis of these findings, the following structure has been proposed for these compounds simulating monomer units in MA—TASM copolymer:

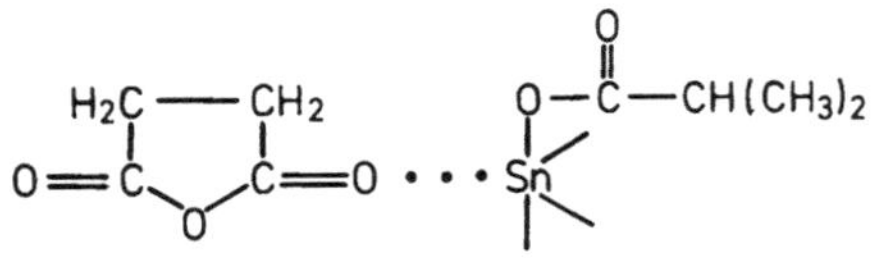

This effect is also evident, although somewhat less pronounced, from the spectra of MA/TASM mixtures which can be ascribed to different electronegativities of the carbon atoms affecting the basicity of neighbouring C=O groups involved in complex formation. The changes in IR-spectra of MA/TBSM and SA/TBMP mixtures suggest the formation of coordination bonds between tin atoms and carbonyl groups of the type:

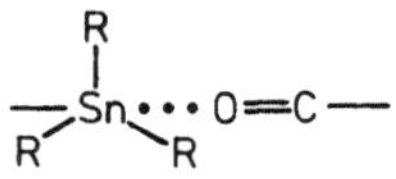

In the spectra of the MA/TBSM mixture, the absorption band of the C=C bond is observed to shift to a shorter wavelength which indicates localization of electron density toward the trialkylstannyl group [2, 19].

The IR-spectra of a mixture of propyl chloride/tributylstannyl 2-methylpropanoate (the saturated analogs of vinyl chloride/TBSM) analogous to vinyl chloride and TBSM units in their copolymer revealed the formation of coordination bonds between polar groups of the —Cl... Sn-type[24, 25].

A shift of stretching vibrations $\nu_{c=o}$ from 1590 to 1640 cm^{-1} is observed in the mixture of model compounds which points to intermolecular coordination interactions.

The H—NMR spectra indicate displacements of chemical shifts for the double bond protons of MA and TBSM and protons of the CH_2—Sn group which suggest that two complex types are present in the mixture of the monomer pairs under study:

$$\text{MA (acceptor)} + \text{TBSM (donor)} \overset{K_c}{\rightleftharpoons} [\text{MA...TBSM}]$$

(Complex formation involving π-electrons of double bonds.)

$$\text{MA (donor)} + \text{TBSM (acceptor)} \overset{K_c'}{\rightleftharpoons} [\text{MA} \ldots \text{TBSM}]$$

(Complex formation involving $-\overset{|}{\underset{/\backslash}{Sn}} \ldots O{=}C$ coordination bonds.)

In the MA/TBSM mixture chemical shifts of double bond protons are displaced to long-wavelength field of MA ($\Delta_{MA} = \delta_{MA}^{fr} - \delta_{MA}^{obs} = 0.025$ ppm) and to the short-wavelength field of TBSM ($\Delta_{TBSM} = \delta_{TBSM}^{obs} - \delta_{TBSM}^{fr} = 0.012$ ppm from CH_2 shift). Excess of MA and equimolar mixtures cause considerable displacement of chemical shifts for the protons of the CH_2—Sn group ($\Delta_{TBSM} = -0.012$ ppm).

To determine the constant (K_c) of complex formation between TBSM and MA from the shifts of protons of anhydride, CH_2 =, and CH_2—Sn-groups, the modified Beneshi-Hildebrand equation [26] was used:

$$\frac{1}{\Delta} = \frac{1}{K_c \cdot \Delta_0 \cdot [D]} + \frac{1}{\Delta_0}$$

where Δ is the difference between the observed shift of acceptor protons in the presence of an electron-donating monomer and the chemical shift of protons in a free electron-donating monomer; Δ_0 is the difference between the proton shift of MA (or TBSM) in MA ... TBSM complexes (charge-transfer complexes [CTC] or coordination complexes) and non-complexed MA (or TBSM); K_c is the complexing constant and [D] the electron-donating monomer concentration.

The results of ^{1}H-NMR studies on a series of solutions of TBSM/MA mixture are summarized in Table 1.

Using the data from Table 1 and plotting $1/\Delta$ vs $1/[\text{TBSM}]$, the value of K_c for the formation of the TBSM ... MA complex was determined (involving π-electrons of double bonds due to the shift of anhydride protons):

$$K_c = 0.03 \pm 0.01 \text{ l/mol (in benzene)} \quad \text{and}$$
$$K_c = 0.04 \pm 0.01 \text{ l/mol (in } CDCl_3) .$$

If [TBSM] $\ll$ [MA] ([MA] = 0.2 mol/l; [TBSM] = 1.6, 3.0, 4.4 and 6.0 mol/l), the following value of the complexing constant was obtained from the shift of olefinic protons in the TBSM group (Δ = 0.01, 0.013, 0.026 and 0.03 ppm) and the plot of $1/\Delta$ vs. $1/[\text{MA}]$:

$$K_c = 0.046 \pm 0.002 \text{ l/mol (in } CDCl_3 \text{ at } 37\ ^\circ C) .$$

This value is in fair agreement with that when considering the displacement of the chemical shifts of anhydride protons.

To determine K_c' for the coordination complex (involving $-\overset{|}{\underset{/\backslash}{Sn}} \ldots O{=}C$ bonds), H-NMR spectra were taken for solutions of TBSM and MA in chloroform in the

Table 1. ^{1}H-NMR results for the determination of the complexing constant (K_c) for TBSM/MA mixtures

[MA] (mol/l	[TBSM] (mol/l)	Δ (ppm)	$\frac{1}{[TBSM]}$	$\frac{1}{\Delta}$	Solvent
0.1	1.0	0.025	1.00	40.0	C_6H_6
0.1	1.5	0.038	0.66	26.6	C_6H_6
0.1	2.0	0.050	0.50	20.0	C_6H_6
0.1	3.0	0.077	0.33	13.0	C_6H_6
0.1	1.0	0.038	1.00	26.6	$CDCl_3$
0.1	1.5	0.053	0.66	19.0	$CDCl_3$
0.1	2.0	0.077	0.50	13.0	$CDCl_3$
0.1	3.0	0.110	0.33	9.0	$CDCl_3$

presence of a high excess of MA ([TBSM] = 0.2 mol/l, [MA] = 10, 6, 4.4 and 3.0 mol/l). K_c' for the formation of the coordination complex was calculated from proton shifts in the CH_2—Sn group and from the plot of 1/Δ vs. 1/[MA]:

$$K_c' = 0.17 \pm 0.02 \text{ l/mol (in } CHCl_3 \text{ at 37 °C)} .$$

Based on a study of the reaction of diethylstannyl dicaprylate with some carboxylic acids (oleic, caprylic, caproic acid, etc.) it was suggested that oleic acid tends to displace caprylic acid from the inner sphere of a carboxylate-type complex forming a sufficiently stable complex with tin. The spectrum of the initial diethylstannyl dicaprylate revealed a band at 1600 cm^{-1} which is intermediate between 1570 and 1720 cm^{-1}. The latter bands are associated with vibrations of C⩦O bonds occurring within an ionized carboxyl group and of C=O bonds in a COOH group, respectively. It was assumed that the band at 1600 cm^{-1} is due to coordination-bound carboxylate ions [27].

The structure of organotin epoxides is characterized by coordination bonds between tin atoms and epoxide oxygen [21,28-30]. The IR spectrum reveals that typical oxirane ring absorption bands of organoepoxystannanes near 3000 and 2985 cm^{-1} shift to 3035–3060 cm^{-1} which points to a higher strain of the epoxy ring. The bands of Sn—C bonds with corresponding frequencies are reported to change in a similar way. Typical oxirane ring bands (3010 and 1100 cm^{-1}) in organotin diepoxides shift to 3055 and 1115 cm^{-1}, respectively. Absorption bands at 450–530 cm^{-1} typical of the normal position of the Sn—C bond [1,31] appear near 610 (ν^{s}_{Sn-C}) and 595 cm^{-1} (ν^{as}_{Sn-C}).

Infrared spectra of unsaturated organotin epoxides show absorption bands at 1235 and 850 cm^{-1}, which are typical of epoxy groups, 3085 and 1640 cm^{-1} (double bonds) and in the range 510–585 cm^{-1} (Sn—C bonds). Symmetrical deformation vibrations of the CH_3 or CH_2 group that is closest to the tin atom are related to changes of H—C—H angles and appear in the 1190–1175 cm^{-1} range [31].

A comparative analysis of the IR spectra of monomers under study and known model compounds (tetraalkylstannanes and unsaturated oxiranes) indicates corresponding shifts of absorption bands for the oxirane ring and the Sn—C bond.

The increase of the strain in the oxirane ring of organotin epoxides can be accounted for by the effect of electron-acceptor groups R_3Sn. It can be assumed that the following coordination interactions between tin atoms and epoxide oxygen occur in the structure of organotin epoxides [28,29]:

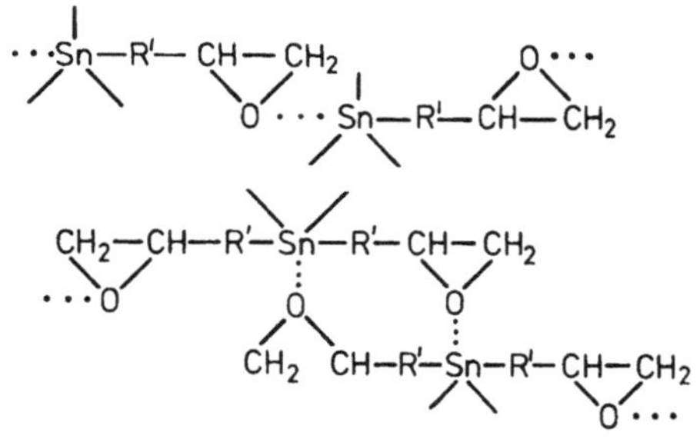

This suggestion is also corroborated by NMR results [21]. From the NMR spectrum of α-trimethylstannylpropenyl glycidyl ether it follows that chemical shifts in the proton signals of the methine and methylene group of the oxirane ring appear in the range of 1.9 to 3.5 ppm as complex splitting with several shifts relative to the normal position ($\delta_{CH_2} = 2.6$ and $\delta_{CH} = 3.1$ ppm), corresponding to the oxirane ring protons in epoxy compounds.

The NMR spectrum of 4,5-epoxy-9-trimethylstannyl-1-nonene consists of six resonance signals [21]; the weak-field low intensity multiplet relates to the protons of the — CH — CH_2 fragment with the chemical shift $\delta = 4.9$ to 5.7 ppm, the protons of the $=CH_2$ group resonating in a stronger field than that of the $=CH$ group. The methylene proton in close proximity to the tin atom appears in a relatively stronger field than CH_2 protons which are observed as stretched signals in the region $\delta = 1.3$ to 1.7 and 2.0 to 2.15 ppm, respectively. The signals of the oxirane protons lie in the region $\delta = 3.1$ to 3.5 (instead of 2.6 to 3.1 ppm for 4,5-epoxy-1,9-nonadiene).

The methylene protons that are very close to the oxirane oxygen atom are deshielded and appear as a broader signal near $\delta = 4.2$. This deshielding appears to be due to both van der Waals interactions between protons and the oxygen atom [33] and the effect of the unshared electron pair at the heteroatom [34].

The strong signal near $\delta = 0.1$ to 0.15 ppm is ascribed to CH_3 protons. In addition, due to spin-spin interactions of CH_3 protons with $^{117/119}Sn$ isotope (natural content 7.67% and 8.67%, respectively), two symmetrically located satellites are observed around their principal signal. The constants of spin-spin interactions ($^{117/119}J_{Sn-C-H}$) in CCl_4 solution are $^{117}J_{Sn-CH} = 51.2$ Hz and $^{119}J_{Sn-CH} = 54.0$ Hz (measurement accuracy ± 0.05 Hz) whereas for the model compound, $(CH_3)_3Sn—CH_2CH_2CH_3$, containing no epoxy group, these values are relatively low, namely 50.4 and 52.6 Hz, respectively [1]. The decrease of the $^{117/119}J_{Sn-CH_3}$ values indicates a coordination-bound state of the atom in the compounds under study. A similar effect of intermolecular interaction and stereochemical non-rigidly of pentacoordinated structures of organotin compounds is reported in Ref. [17].

A certain interaction between tin and chlorine atoms through π- and σ-electron systems has been confirmed by spectral structural studies of chloroorganotin adducts and model compounds [35].

A noticeable change in the IR spectra of trialkylstannyl chloroendiconates occurs in the region 1500–1900 cm^{-1}. In this case the stretching vibration of the C=O group appears in various regions, depending on the nature of substituents at the ô-carbon atom (with respect to C=O) and the length of the alkyl chain at the tin atom. The introduction of chlorine into the ring shifts the C=O absorption bands to 1730 cm^{-1} (chlorendic anhydride) and 1670 cm^{-1} (bis-tributylstannyl chlorendiconate), respectively. Hence, conjugation between chlorine and tin through the π- or σ-electron system occurs, thus enhancing the covalent nature of the C—Sn—O bond and the stability of the latter to hydrolytic decomposition.

When investigating H-NMR spectra of bis-trialkylstannyl chloroendiconates, the chemical shift of symmetrical ring protons caused by interactions between chlorine and carbonyl atoms was utilized analytically. A comparative analysis of H-NMR spectra of model compounds (endic and chlorendic anhydrides and their derivatives) and bis-trialkylstannyl chlorendiconates reveals that the chemical shift of the CH group of the ring is significantly affected by the nature of the surrounding substituents and by the distant alkyl groups at the tin atom. The introduction of chlorine into the molecule results in a considerable displacement ($\Delta\delta = 0.84$ ppm) of the CH chemical shift to a weaker field. A shift of δ_{CH} is also observed with increasing length of the alkyl chain at the tin atom which is due to improved electron-acceptor properties of the R_3Sn group coordinated with a carbonyl group [35]:

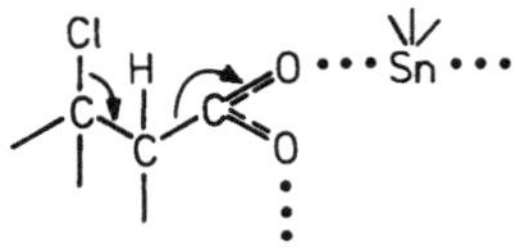

X-ray crystal analysis of certain organometallic compounds of the silicon group by Gusev et al. [36] showed that tin and chlorine (or nitrogen) atoms in some organotin compounds are brought together to a distance shorter than the sum of the van der Waals radii which points to a coordination interaction between these atoms [35]:

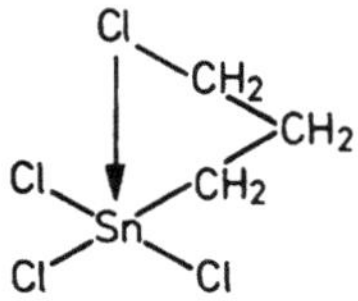

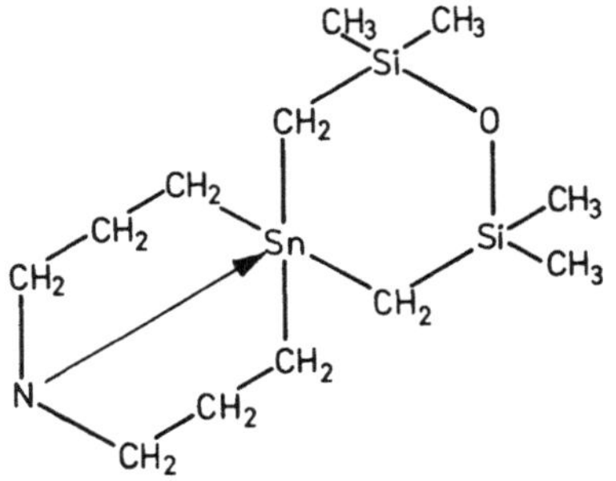

The distance of Sn ... Cl and Sn ... N in these compounds are 3.28 A and 2.36 A, respectively.

Kuzmina and Struchkov [37] reported the structure of some organic compounds of transition metals, including triphenylstannyl derivatives of 2-dimethylaminothiophenol and 8-mercaptoquinoline. X-ray analysis revealed that these molecules contain secondary intramolecular bonds

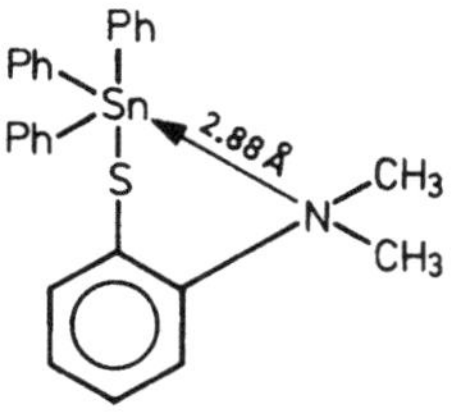

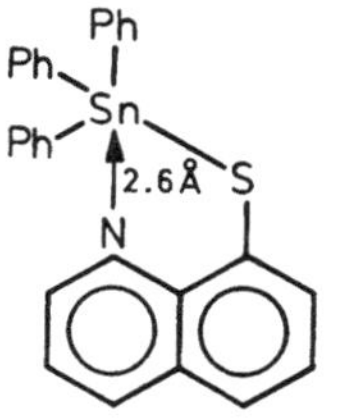

which exert a marked effect on the course of some metal-metal exchange reactions (Sn—Hg).

3 Polymerizations and Copolymerizations

Among the great variety of organotin compounds those with polymerizable groups are of paramount importance.

Depending on the type of conjugation and the mutual location of the tin atom and reactive groups, all organotin monomers can be categorized into the following major groups: monomers with $d_\pi - p_\pi$ conjugation (vinylstannanes, acetylene and mixed vinyl-acetylene derivatives, conjugated tin-containing dienes and trienes, etc.); monomers with d_π-σ-π conjugation (allyl, epoxypropyl monomers, etc.); monomers containing electron-rich groupings (or atoms) located between tin atoms and polymerizable groups (organotin derivatives of unsaturated acids and vinyl-aromatic compounds, i.e. monomers with $d_\pi - p_\pi$ (heteroatom) conjugation, etc.).

This review is mainly concerned with monomers of the latter group which exhibit the highest reactivity towards polymerization and copolymerization reactions.

As far back as 1937, Andrianov reported that polymeric compounds containing group IV B elements were of great theoretical interest and could acquire far-reaching importance. Andrianov and co-workers carried out fundamental research on polymers with inorganic main chains and made a valuable contribution to the chemistry of heteroorganic polymers. They also marked the beginning of a new direction in this field, the synthesis of high-molecular weight compounds with alternating silicon, oxygen and metals (tin, titanium, aluminium, boron, etc.) [38-41].

A great number of organotin compounds containing polymerizable functional groups have been described in the literature [42-47].

It is commonly known that vinyltin compounds undergo no free-radical polymerization and do not readily copolymerize with various vinyl monomers which can be attributed to their inhibiting effect towards radical reactions [48,79].

Free-radical copolymerization of trimethyl- or tributylvinyltin with styrene or methyl methacrylate results in low (~10%) yield of copolymer. Moreover, both the reaction rate and viscosity decrease considerably with higher vinyltin content in the starting mixture [49]. These findings imply that organotin monomers tend to inhibit free-radical copolymerization.

However, vinyltin monomers are readily polymerized or copolymerized with butyllithium at 0 and 20 °C respectively by an anionic mechanism [51].

The absolute reactivity of vinalstannanes was evaluated by copolymerizing them in bulk with ethylene at 160 °C and 1400 kg/cm^2 in the presence of dibuthyl peroxide. In the system triethyl-vinylstannane/ethylene, as in other systems, the reactivity of both vinylstannane and its radical is found to decrease due to conjugation within the molecule. The mean effective copolymerization constants are $r_1 \sim 0$ and $r_2 = 3.5 \pm 1.0$ [52].

Andrianov and Zhdanov have developed a method for the synthesis of polymers containing heterochain and carbon-chain units by free-radical copolymerization of metal-containing polyorganosiloxanes bearing a pendant vinyl group with vinyl monomers. The copolymers thus obtained display increased thermal stability and can be used for the production of laminated plastics, adhesives and other valuable materials [53].

It is interesting that the inhibiting action of unsaturated organotin compounds is strongly dictated by the arrangement of the double bond within the molecules. When the vinyl group is directly connected with the tin atom, the inhibiting effect is strong but when it is joined via polar groups such as phenyl, carboxy, etc. to the tin atom, the inhibiting action of tin compounds as considerably weaker. Thus, organotin acrylates, methacrylates or styrene derivatives can be readily polymerized or copolymerized with various monomers.

To improve the polymerizability of organotin monomers, a convenient method for the production of trialkylstannyl-1,3-alkadienes has been developed [54–57] and their copolymerization with styrene and methyl methacrylate studied [58,59].

Along with low-molecular weight organotin compounds, which have found extensive application as plastics stabilizers and biocides, antioxidants, catalysts, and physiologically active substances, carbon-chain polymeric tin compounds have found growing interest. The reason for this is that the introduction of trialkylstannyl groups into a macromolecular chain results in polymer or copolymer products acquiring the above specific properties typical of organotin compounds.

It is believed that novel film-forming materials for coatings will be based on organometallic polymers, the food prospects for organotin film-forming polymers being particularly emphasized [60].

Recent survey articles [61,62] are mainly concerned with the synthesis, properties and applications of carbon-chain organotin polymers and with the use of organotin compounds in polymer chemistry as stabilizers, fungicides, etc. It should particularly be noted that polymers on the basis of trialkylstannyl methacrylates exhibit biocide properties.

Trialkyl (triaryl)stannyl methacrylates were copolymerized with ethylene and methyl methacrylate and it was shown that the resulting copolymer offers improved mechanical properties as compared to ethylene, and high fungicidal activity. It was suggested that homopolymers and copolymers of triethylstannyl methacrylate contain a covalent and an ionic bond between the carboxy group and the tin atom [63].

A new method has been proposed for obtaining organotin acrylates and methacrylates by a bulk reaction of the corresponding acids in stoichiometric quantities with trialkyl(triaryl)tin hydroxides and hexaalkyldistannoxanes in the presence of inert dehydrating agents ($MgSO_4$, or $CaSO_4$) and p-methoxyphenol as polymerization inhibitor. The highest yield obtained is about 90%[64].

The authors of Ref.[65] prepared a number of organotin maleates and were the first to carry out diene syntheses using the maleates

$$C_6H_5OOC{-}CH{=}CH{-}COOSnR_3 \quad \text{or}$$
$$(n\text{-}C_4H_9)_2Sn(OOC{-}CH{=}CH{-}COOCH_3)_2$$

as a dienophil and 2,3-dimethyl-1,3-butadiene or cyclopentadiene as a diene monomer.

The synthesis of organotin derivatives of itaconic and citraconic acid by the known reaction of trialkyl(triaryl)stannanes with the corresponding unsaturated acids has been reported[66]. These acids are presumed to be of interest for the production of highly bactericidal film-forming polymers.

The synthesis of organotin "oligosteracrylate" i.e. dimethylstannyl dimethacrylate, and the production of the cross-linked homopolymers on its basis have been reported. Morphology, mechanical and relaxation properties of poly(dimethylstannyl dimethacrylate) have been investigated[67].

Water-repellent coatings for the protection of glass, plastic or metal surfaces can be obtained on the basis of a stannosiloxane chelate (reaction product from tributylchlorostannane and methyl octadecylchlorosilane)[68].

Organotin carboxylates ($R_3SnOOC{-}R'$) are applied to metal surface as a thin layer and hardened by heating to high temperatures (about 400 °C)[69].

Epoxytin compounds are obtained by either hydrostannylation of unsaturated epoxides or condensation of organotin alcohols with epichlorohydrin[70]. Extensive applications of these compounds in plastic industry have been suggested.

The introduction of organotin residues into an epoxide resin is known to improve the dielectric properties and thermal stability of hardened compositions[71].

A Japanese patent[72] claims the synthesis of thermally stable copolymers by free-radical terpolymerization of dialkylstannyl dimethacrylates, glycidyl methacrylate and vinyl monomers (vinyl chloride, styrene, vinyl acetate, etc.). The products contain 0.5 to 30% tin and 0.05 to 7% epoxide oxygen.

Organotin compounds with the general formulae $RSnX_3$, R_2SnX_2, R_3SnX and RSnOOH (R = alkyl or aryl and X = Cl, Br or J) are efficient catalysts for the polymerization of oxiranes[73].

Copolymerization of tri-n-butylstannyl acrylate and methacrylate with vinyl monomers containing epoxy and hydroxy groups results in the formation of biologically active organotin copolymers which can be hardened with aliphaticand aromatic amines. The autocatalytic effect of organotin groups has been revealed in self-cross-linking of the copolymer[74].

Copolymerization constante and the parameters Q and e of these monomers have also been determined[75]. The highest tendency for alternating copolymerization exhibits the tri-n-butylstannyl methacrylate/methacrylate system. Based on the distribution of comonomer units, appropriate monomer systems have been proposed for obtaining thermosetting organotin coatings with favorable mechanical and fungicidal properties.

Copolymers of tri-n-butylstannyl acrylate with alkyl acrylates (methyl, athyl, butyl or octyl acrylates) or acrylonitrile can be obtained in solution or in a slurry at 72 °C. Ethyl, butyl and octyl acrylates display a tendency to form alternating monomer units in these systems. The rate of copolymerization decreases with increasing tri-n-butylstannyl methacrylate concentration in the starting monomer mixture. Free-radical copolymerization of the same monomer pairs was carried out in DMFA solution at 50 °C. The copolymerization constants differ slightly from those found for the emulsion copolymerization [76–78].

Free-radical multicomponent copolymerization of dialkylstannyl maleates or dialkylstannyl dimethacrylates with methallyl alcohol (or β-hydroxyalkyl acrylates) and vinyl monomers (sryrene, methacrylic acid or methacrylamide) yields polymeric powders. Due to their storage and thermal stability and impact strength they are used as protective coatings [79].

Yamada et al. [80] investigated the free-radical homopolymerization of dimethylstannyl dimethacrylate (DMSM) and its copolymerization with trimethylstannyl methacrylate (TMSM). The absorption bands at 527 and 513 cm^{-1} found in the IR spectra of the monomers were assigned to dimethylstannyl (R_2Sn^{2+}) and trimethyl stannyl (R_3Sn^{+}) cations, respectively. The TBSM homopolymer was hydrolyzed and then methylated with diazomethane to yield poly(methyl methacrylate). It was shown by IR studies that TBSM homopolymer exhibits a higher content of syndiotactic triads than poly(methyl methacrylate) synthesized by radical polymerization. Yamada et al. have compared DMSM and TMSM polymerization with the free-radical polymerization of methyl methacrylate in the presence of $SnCl_4$ as a complexing agent.

Organotin polymers containing one carboxyalkyl group bound to the tin atom in each recurring monomer unit are obtained by polymerization of an organotin acid of the type $Sn[(CH_2)_mCOOH]_4$ in aqueous medium at 60 °C. The powdery products are used as biocides and stabilizers for synthetic polymers [81].

Tin-containing polymers obtained by polymerization of ethylene with tetraethyl- or tetrabutylstannane at 1400 kg/cm^2 and 160–200 °C were used as antiwear additives for lube oils [82].

The enthalpy of polymerization of tri-n-butylstannyl methacrylate ($\Delta H_p^\circ = 60.5 \pm \pm 1.2$ kJ/mol) has been measured in an isometric calorimeter [83].

A scientific collection [84] of the large number of data on organometallic polymers containing various elements has been published. Tin-containing polymers, their use and their applications as coating materials are also discussed in this collection.

In recent reviews [85,86] are discussed the uses of organotin compounds as stabilizers and catalysts in polymer chemistry.

3.1 Effects of Intermolecular Coordination

Trialkylstannyl methacrylates (TASM) readily undergo free-radical copolymerization with maleic anhydride [2,19], styrene [87], and vinyl chloride [24,88].

Of considerable interest are the findings on the free-radical copolymerization of TASM with maleic anhydride (MA).

Based on the structure of the monomers in the monomer feed, the systems under study can be classified as acceptor (MA) — acceptor (TASM) systems and the assumption that alternating copolymerization occurs in these systems seems at first sight to be alluring However, detailed investigations of the effects observed in free-radical copolymerization due to the presence of trialkylstannyl groups in the polymerizing monomers as compared to the corresponding organic analogs confirmed the presence of $-\overset{\diagdown\,\diagup}{\underset{|}{Sn}}\ldots O{=}C-$ coordination bonds in these systems. This intermolecular coordination occurring in free-radical copolymerization of MA with TASM is thus responsible for monomer unit alternation along the macromolecular chain.

Inspection of Table 2 shows that the tendency for alternation of the monomer units grows with the length of alkyl substituents at the tin atom in organotin methacrylates. This results in changes of the specific activity (Q_2) and polarity (e_2). A decrease in Q_2 of organotin methacrylates as compared to alkyl acrylates indicates the effect of the electron-accepting groups SnR_3 on the conjugation between the C=C bond and the carbonyl group.

Substitution of a tert-butyl group in tert-butyl methacrylate, $(CH_3)_3COOC{-}C(CH_3){=}CH_2$, by a trimethylstannyl group yielding $(CH_3)_3Sn{-}OOC{-}C(CH_3){=}CH_2$ results in a decrease of Q_2 from 1.18 to 0.31 and an increase of e_2 from 0.35 to 0.64.

These results confirm that the tendency for an alternating arrangement of the comonomer units increases with the length of alkyl chains at the tin atom.

It is known that the effect of a substituent in a radical is significantly different from that in a monomer. Thus, trialkylstannyl groups in TASM exhibit a different propensity for coordination interactions with the carbonyl group of MA, the acceptor properties of the SnR_3 groups becoming stronger when they are incorporated into the MA molecule or in the interaction with MA. The same applies when conjugation effects in TASM and MA disappear (in the saturated analogs or in the side-chains of the macromolecules). Trialkylstannyl grups are responsible for the reverse conjugation between the carbonyl group and the double bond, thus preventing electron delocalization.

This assumption is confirmed by the negative e_2 values. Organic analogs of TASM exhibit positive polarities [89].

The specific activity (Q_2) of TASM falls from 0.31 to 0.18 with increasing length of the alkyl substituent, i.e. with growing effect of spatial factors. The polarity (e_2) changes in the reverse direction (from —0.88 to —0.64) whereas in the organic analogs e_2 decreases with rising length of the comonomer substituents, due to the depolarizing effect of the non-polar alkyl residues on the C=C double bond.

Table 2. Copolymerization constants (r_1 and r_2), specific activities (Q_2) and polarities (e_2) for MA — TASM pairs [2, 19]

Monomer paris	According to Finemann-Ross					According to Jaks				
	r_1	r_2	$r_1 \cdot r_2$	Q_2 [a]	e_2 [a]	$r_1 \times 10^3$	r_2	$r_1 \times r_2 \times 10^4$	Q_2 [a]	e_2 [a]
MA — TMSM	0	0.220	0	0.31	—0.64	0.8	0.18	1.4	0.35	—0.72
MA — TESM	0	0.122	0	0.27	—0.75	0.6	0.11	0.66	0.34	—0.85
MA — TPSM	0	0.081	0	0.23	—0.81	0.6	0.08	0.48	1.31	—0.90
MA — TBSM	0	0.053	0	0.18	—0.88	0.8	0.05	0.4	0.26	—0.92

[a] Calculated for reported values $Q_1 = 0.23$ and $e_1 = 2.25$ for MA [89].

From the r_1 and r_2 values it may be inferred that trialkylstannyl electron-accepting groups suppress almost completely the addition of TASM to its own radical.

By comparing the constants of the copolymerization of MA with organotin methacrylates with the known values for the copolymerization of the MA — MMA system ($r_1 = 0.03$ and $r_2 = 3.5$) [89] where almost no complexation takes place, the following conclusion on the effect of the electron-accepting groups SnR_3 on free-radical copolymerization and be reached.

The reaction proceeds at the stage of pseudo-cyclocopolymerization involving complex formation between the growing macroradical and the monomer which is responsible for the alternation of monomer units along the macromolecular chain:

$$\mathrm{MA} + \mathrm{TASM} \underset{\text{(complexation)}}{\overset{K_c}{\rightleftharpoons}} \mathrm{MA}\cdots\mathrm{TASM} \tag{1}$$

$$\mathrm{R}^{\bullet} + \mathrm{MA}\cdots\mathrm{TASM} \xrightarrow[\text{(initiation)}]{} \mathrm{R{-}MA}^{\bullet}\cdots\mathrm{TASM} \tag{2}$$

$$\mathrm{R{-}MA}^{\bullet}\cdots\mathrm{TASM} \xrightarrow[\text{(intramolecular growth)}]{} \mathrm{R{-}MA{-}TASM}^{\bullet} \tag{3}$$

$$\mathrm{R{-}MA{-}TASM}^{\bullet} + \mathrm{MA} \rightarrow \mathrm{R{-}MA{-}TASM}^{\bullet}\cdots\mathrm{MA} \rightarrow \mathrm{R{-}MA{-}TASM{-}MA}^{\bullet} \quad \text{(intermolecular growth)} \tag{4}$$

According to this scheme the fundamental difference in the mechanism of free-radical copolymerization of MA with TASM and of MA with alkyl acrylates is due to the fact that in the former copolymerization intermolecular coordination is involved. This coordination is similar to the effect of various complexing agents ($ZnCl_2$, $SnCl_4$ and $AlCl_3$) on free-radical homo- and copolymerization of vinyl monomers. This effect seems to favor the appearance of isotactic configurations along the main chain.

It is known that the polymerization of MMA ... $SnCl_4$ and $(MMA)_2$ ... $SnCl_4$ complexes (MMA = methyl methacrylate) yields a polymer which predominantly exhibits an isotactic structure [90]. From the analogy between these complexes and those discovered by the author of this article (MA ... SnR_3), it can be suggested that free-radical copolymerization of MA with trialkylstannyl methacrylates yields copolymers mainly exhibiting configuration [19].

It is likely that the observed coordination interaction between individual segments of the macroradicals and monomer units determines the stereoorientation of free-radical copolymerization of organotin methacrylates with MA.

As indicated above, the observed effect of intermolecular coordination results in the formation of alternating copolymers having coordination bonds between fragments of the corresponding comonomers.

Along with spectroscopic techniques, the presence of $-\overset{|}{\mathrm{Sn}}\ldots\mathrm{O{=}C}\!<$ in the copolymers studied was confirmed by viscosimetric data [2].

The dependence of intrinsic viscosity on concentration has a non-linear character, which is probably due to the conformational changew caused by the destruction of

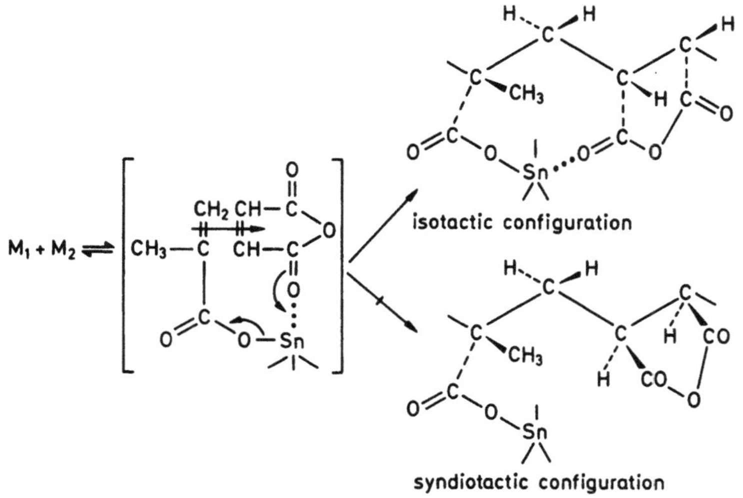

intermolecular coordination bonds by DMFA which displays a strongly solvating effect $(CH_3)_3\ N^+ {=\!=} C {=\!=} O^-)$.

A fairly clear-out linear dependence $\eta_{sp}/C \rightarrow C$ is observed for organotin copolymers dissolved in DMFA in the presence of 0.1% LiCl and for DMFA solutions of the organic analogs of these copolymers [2].

Viscosimetric studies of organotin copolymer solutions allow the changes in the shape of the macromolecules to be followed as a function of the electrostatic charge. From the plot of the intrinsic viscosity of copolymers in DMFA solution against the degree of dilution it is seen that increasing dilution results in a rise of viscosity, probably due to an extension of macromolecular chains accompanied by conformational transformations. Naturally, this rise in viscosity with dilution cannot proceed infinitely since a coiled chain cannot be extended more than a completely extended chain conformation, due to intramolecular repulsion.

An anomalojs character of the dependence of the viscosity on dilution may be explained by the course of conformational changes before a stable extended chain conformation has been formed after which the linearity of the viscosity-dilution relationship is restored.

The observed anomaly in the viscous properties of dilute organotin copolymer solutions seems to be attributable to the existence of both intra- and intermolecular associates,due to coordination interactions between SnR_3 and $C{=}O$ fragments of side groups.

The $\eta_{sp}/C \rightarrow C$ curves become straightened when the ionic strength of the solution is increased by the addition of 0.1% LiCl. The addition of a strong electrolyte (LiCl) to a DMFA solution of copolymers apparently results in a extension of the macromolecular chains and partial liberation of the associated fragments from electrostatic attraction of opposite charges (R_3Sn^+ and $^-O{=}C$).

These empirical observations confirm that primary supramolecular associates typical of conventional polyelectrolytes rather than individual intermolecular interactions exist in organotin polymer solutions.

It can be assumed that a similar ligand destruction takes place in the starting monomer mixture under the action of nucleophilic solvents. In fact, it was found that MA practically fails to copolymerize with organotin methacrylates in DMFA (dielectric constant $\varepsilon = 37$), and the monomeric system acquires a strong coloration.

Dissolution of a MA/TBSM mixture in DMFA leads to dissociation of the complexes and hence shifts the reaction towards the formation of random copolymers or the propagation reaction is inhibited by the competing solvatation effect of the solvent.

To elucidate the reaction mechanism, the kinetics of free-radical copolymerization of the monomers concerned was investigated.

If monomer conversion is high, the linear time dependence of conversion is retained and the initial copolymerization rate decreases with increasing length of the alkyl substituent at the tin atom in trialkylstannyl methacrylates, most probably due to their enhanced ability to undergo coordination interactions with the carbonyl groups. The considerable increase in the total rate of copolymerization of MA and trialkylstannyl methacrylates ($0.88 \times 10^{-5} - 3.3 \times 10^{-5}$ mol/l s) as compared to that of MA and n-butyl methacrylate (BMA) (0.4×10^{-5} mol/l s) can be explained by the higher cross propagation rate. Moreover, the increased number of side groups and coordination bonds between the comonomer units may prevent radical interactions and favor kinetic chain propagation [2,19].

A similar change is observed when comparing homopolymerization rates of TBSM and BMA. The polymerization rate of TBSM (6.66×10^{-5} mol/l s) is higher by a factor of 4.3 than that of its organic analog, BMA (1.54×10^{-5} mol/l s).

In all monomer systems studied a maximum polymerization rate was observed at an equimolar ratio of the starting monomers when the probability of the formation of the complex and its concentration are higher than for other ratios.

At an equimolar ratio of the starting monomers the polymers formed have the highest intrinsic viscosity.

Alternating copolymerization of trialkylstannyl methacrylates with MA can proceed via several routes, e.g. by sequential addition of free monomers to the macroradical, by the addition of monomer pairs to a complex or with simultaneous contribution of both free and complex-bound monomers.

To elucidate the chain propagation mechanism of alternating copolymerization of TBSM with MA and to quantitatively estimate the contribution of complex-bound monomers to the propagation reaction, a kinetic approach described in Ref. [91] was employed.

First, it was necessary to determine the position of the maxima on the plots of TBSM — MA copolymerization rate vs. composition of the monomer mixture at various total monomer concentrations. The observed shift of the maxima of the rate as a function of the dilution of the monomer mixture toward higher MA concentrations is a consequence of a complex mechanism, i. e. both free and complex-bound monomers are involved in chain propagation reactions:

$$\sim TBSM^{\cdot} + MA \xrightarrow{K_{12}} \sim TBSM{-}MA^{\cdot} \tag{5}$$

$$\sim TBSM^{\cdot} + MA \ldots TBSM \xrightarrow{K_{1C}} \sim TBSM{-}MA{-}TBSM^{\cdot} \tag{6}$$

$$\sim MA^{\cdot} + TBSM \xrightarrow{K_{21}} \sim MA - TBSM^{\cdot} \tag{7}$$

$$\sim MA^{\cdot} + TBSM \ldots MA \xrightarrow{K_{2C}} \sim MA - TBSM - MA^{\cdot} \tag{8}$$

Proceeding from the copolymerization rate equation

$$v = -\frac{d([A]+[D])}{dt} = K_{AD}[\sim A^{\cdot}][D] + 2K_{AC}[\sim A^{\cdot}][C] + K_{DA}[\sim D^{\cdot}][A] + 2K_{DC}[\sim D^{\cdot}][C] \quad (9)$$

with the condition of maxima dv/dA = 0 at [M] = const, an equation was obtained which allows to determine α, β_1 and β_2 from several v_{max}:

$$a\alpha\beta_1 + b\alpha\beta_2 + c\alpha + d\beta_1 + e\beta_2 = f \quad (10)$$

where

$$\alpha = \frac{K_{12}}{K_{21}} = \frac{K_{AD}}{K_{DA}}, \qquad \beta_1 = \frac{K_{2C}}{K_{21}} = \frac{K_{AC}}{K_{AD}}, \qquad \beta_2 = \frac{K_{1C}}{K_{12}} = \frac{K_{DC}}{K_{DA}}$$

with A = MA, D = TBSM (complex formation proceeds through $-\overset{|}{\underset{/\backslash}{Sn}}\ldots O{=}C$ bonds; K_c = 0.17 l/mol at 37 °C).

$$a = 2\,K_c[A]\,[D]\,, \qquad b = K_c([D]^2 - [A]\,[D])\,,$$

$$c = [D], \qquad d = \frac{K_c[A]^2\,([D]+[A])}{[D]},$$

$$e = -2\,K_c[A]^2\,, \qquad f = [A]^2/[D]$$

Based on experimental data and using the above sets of equations, the following β_1, β_2 and α values were found:

$$\beta_1 = K_{2C}/K_{21} = 2.26\,; \qquad \beta_2 = K_{1C}/K_{12} = 2.11\,;$$
$$\alpha = K_{21}/K_{12} = 0.873\,.$$

The obtained value of α indicates the proximity of the rate constant values of the addition of TBSM to the macroradicals $\sim MA^{\cdot}$ and of MA to $\sim TBSM^{\cdot}$. This can be explained by a similar influence of intermolecular coordination on chain propagation. The values of β_1 and β_2 indicate that in free-radical copolymerization of TBSM with MA both free and complex-bound monomers are involved in chain propagation with a higher contribution of the latter.

It has been suggested that the formation of a coordination complex between comonomers and growing radical species reduces the reactivity of the TBSM ... MA complex towards growing $\sim MA^{\cdot}$ and $\sim TBSM^{\cdot}$ radicals.

Binary free-radical copolymerizations of organotin derivatives of unsaturated acids (tri-n-butylstannyl methacrylate, bis-triethylstannyl maleate (TESM) and β-phenyl-tri-n-butylstannyl methacrylate (PBSM)) with certain vinyl monomers such as styrene (St) [87] and vinyl chloride (VC) have been studied [24,25,92].

A noticeable change in the TBSM activity as compared with that of its organic analog BMA ($r_{BMA} = 0.64$ and $r_{St} = 0.54$)[89] in the free-radical copolymerization with styrene may be ascribed to steric factors and the effect of intermolecular coordination.

Free-radical copolymerization of organotin derivatives of unsaturated acids such as bis-TESM, TBSM and PBSA with vinyl chloride has been carried out[24,25,92].

The values of K and β($K > 0$ and $0 < \beta < 1$) were calculated for each monomer pair from the logarithmic plot of the ratio of the monomers in the monomer feed $[M_1]/[M_2]$ to the comonomer units in the copolymer using a modified equation of binary copolymerization:

$$d[M_1]/d[M_2] = K([m_1]/[m_2])^{\beta}$$

The resulting values point to the fact that organotin monomer units enter the macromolecular chain. To reveal the contribution of trialkylstannyl groups to radical copolymerization, the copolymerization of their organic analogs (BMA and MA) with VC was investigated.

From the experimental results at low conversions (≦10%), copolymerization constants, specific activities (Q) and polarities (e) were determined for the monomer pairs under study. The values obtained were as follows:

$r_{bis\text{-}TBSM} = 0.001$

$r_{VC} = 0.85$, $Q_1 = 0.016$, $e_1 = -2.44$;

$r_{PBSA} = 0.026$, $r_{VC} = 0.057$, $Q_1 = 0.23$, $e_1 = -2.35$;

$r_{TBSM} = 4.5$, $r_{VC} = 0.45$, $Q_1 = 0.20$, $e_1 = -0.71$.

The specific reactivity of organotin monomers toward VC decreases as follows: bis-TESM > PBSA > TBSM[25].

A considerably higher activity of organotin monomers as compared to their organic analogs appears to be due to the complexing effect between the tin and chlorine atoms.

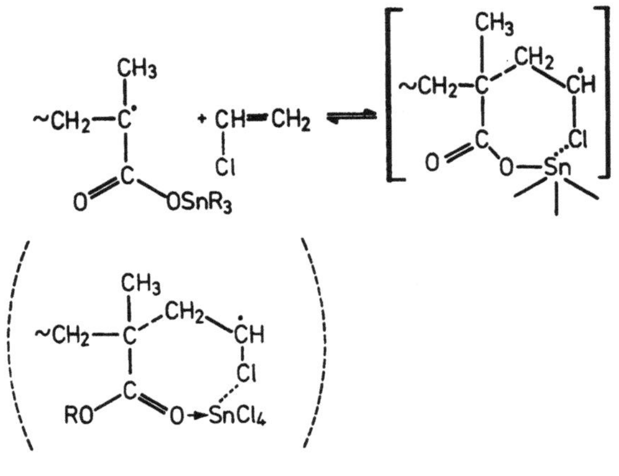

It is seen from kinetic studies of tributylstannyl methacrylate polymerization initiated by azodiisobutyonitrile that the value of K_p/K_0^2 varies between 0.38 and 0.57, depending on the intiating efficiency, and is far greater than for the polymerization of alkyl methacrylates. At high radical concentrations the reaction order with respect to initiator concentration becomes lower than 0.5 which can be explained by the ease of amcrochain termination by primary radicals. This behaviour of tri-butylstannyl methacrylate may be ascribed to a stronger tendency of the tin atom for coordination interactions involving the disappearance of its conjugation effect with π-electrons of the C=O double bonds during chain propagation [93].

When studying the free-radical copolymerization of methacrylic and acrylic acids with vinyl monomers, it was established that the addition of catalytic amounts of $SnCl_4$ and $(C_6H_5)_3SnH$ has a marked effect on the copolymer composition. It was found that complexes are formed by charge transfer between unsaturated acids and the above tin compounds. It has been suggested that the change in polymer composition is caused by the interaction of the tin compounds with a transition complex resulting in a decrease of the resonance stabilization of the latter [94].

Some characteristics of free-radical terpolymerization of tri-butylstannyl methacrylate, styrene and maleic anhydride governed by the pentacoordination state of the tin atom are reported in Refs. [95],[96]. It is shown that a coordination-bound monomer has a considerable effect on chain initiation and propagation. Copolymerization mainly involves the participation of complex-bound monomers.

3.2 Polyaddition Reactions

The method of migration polymerization (polyaddition reaction) finds extensive application in the production of silicon-, germanium- and tin-containing heteroorganic polymers [97].

This method was first applied to the synthesis of organotin epoxy oligomers [98,99].

Some mechanisms of radical-initiated migration copolymerization of dialkyl(diphenyl)stannanes with non-conjugated epoxyalkadienes such as 4,4-epoxy-1,7-heptadiene (I) and 3-glycidyl-oxy-1,6-hexadiene (II) have been discussed [98].

The reaction proceeds as follows:

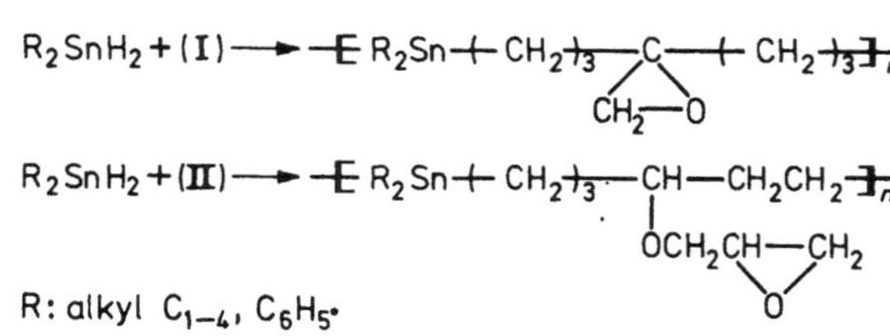

R: alkyl C_{1-4}, C_6H_5.

The mean rate constant at 60 °C ($K \times 10^7$) of the reaction of diphenylstannane with (I) is 6.5 $l^2/mol \times s$ whereas for the reaction of dialkylstannanes with (I) it varies from 0.7 to 1.78 $l^2/mol \times s$. It follows from these findings that diphenylstannane is a more active comonomer than its dialkyl analog.

By studying the initial copolymerization rate as a function of the initial concentrations of monomer and initiator(azo-bis-isobutyric acid, (I)) it was found that the order of di-n-butylstannane with respect to (I) is 0.5 for the monomers and 2.46 with respect to the initiator. On the basis of these data, the copolymerization rate equation can be expressed as follows:

$$V = K[I]^{0.5}[M]^{2.46}$$

The activation energy (E_a) of the migration copolymerization of $(n\text{-}C_4H_9)_2SnH_2$ with (I) calculated from the plot of the rate constant vs. temperature is 12.2 kcal/mol.

The rate of copolymerization is markedly affected by the character of the substituents at the tin atom in organostannanes. The highest reaction rate is obtained at equimolar ratio of the starting monomers. The reactivity of diorganostannanes increases with rising length of the alkyl substituents. A maximum reaction rate is characteristic of copolymerizations involving diphenylstannane. These findings permit to establish the following order of increasing reactivity of organostannanes, depending on the substituent at the tin atom [98].

$$CH_3 < C_2H_5 < n\text{-}C_3H_7 < n\text{-}C_4H_9 \leqq C_6H_5$$

These results combined with the total suppression of copolymerization in the presence of hydroquinone as inhibitor indicate that hydrostannylation takes place upon the polyaddition of diorganostannane to the epoxyolefine by a radical mechanism accompanied by hydrogen atom migration in each chain propagation, No addition of organostannanes to the oxirane ring was observed [98].

3.3 Some Specific Features of the Polymerization of Organoepoxystannanes

Organotin epoxide monomers [30, 100–104] containing fairly reactive oxirane rings and both C=C and Sn—C bonds can be used as starting components for the synthesis of new polymers, chemically active stabilizers and biocides for polymeric materials, e.g. for PVC.

The investigation of the mechanism of organoepoxypolystannane formation shows that radical hydrostannylation of epoxides proceeds mainly during the addition of organostannanes to the C=C bond according to the Former rule, the oxirane ring remaining unaffected [100, 102].

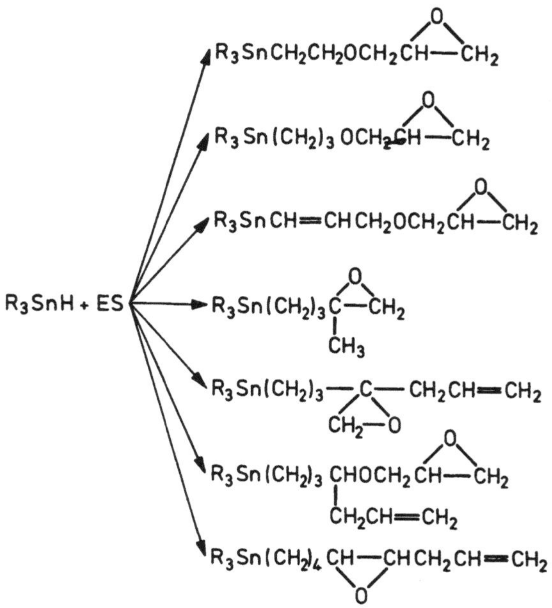

where R = C_{1-4} alkyl; ES = epoxy compounds such as vinyl, allyl and propargyl glycidyl ethers, 4-methyl-4,5-epoxy-1-pentene, 4,4-epoxy-1,7-heptadiene, 3-glycidyl-oxy-1,6-hexadiene and 4,5-epoxy-1,9-nonadiene.

If dialkylstannanes are employed in this reaction, α, ω-diepoxyorganostannanes are formed [30]:

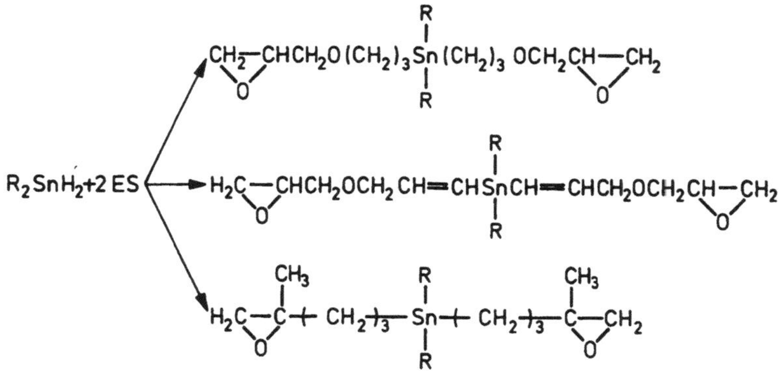

In contrast to α, ω-diepoxyorganostannanes, trialkylstannyls are not readily polymerized. When stored in the free air, they are gradually converted to a powder which may be explained by opening of the oxirane ring upon the catalytic action of alkyl(aryl)stannyl groups.

This effect is more clearly revealed in the polymerization of α, ω-diepoxyorganostannanes [29].

Bulk polymerization of di-n-butyl-bis-(γ-glycidyloxypropyl)stannane in the air or in benzene as reaction medium at 30 °C results in a gradual increase in viscosity and precipitation of a white powder in quantitative yield. The polymerization product is soluble in polar organic solvents and hot aromatics which indicates a linear structure of organotin macromolecules [29].

The spectra of the samples taken from the reaction medium at different times revealed a decreased intensity of the absorption bands near 1250, 1145 and 965 cm^{-1} which are typical of epoxy groups.

Since spontaneous polymerization is only observed in contact with air or in undesiccated benzene used as a reaction medium, it is natural to attribute this phenomenon to the catalytic effect of traces of water on the opening of the oxirane ring involved in the coordination interaction with the tin atom.

The following evidence confirms this concept:

1) addition of small amounts of water to the polymerization medium results in immediate polymerization until the whole viscous mass has been converted to a powder-like product;
2) a similar effect is not observed in organic and organosilicon analogs and
3) dialkyltin derivatives exert a catalytic effect on the polymerization of epoxy compounds [105].

It is known that organotin compounds such as n-butyltin acid and di-n-butyltin dilaureate accelerate the hardening process of epoxy resins [106] and initiate ethylene oxide polymerization [107].

The above considerations stimulated investigations of the polymerization of model systems, namely ethylene oxide in the presence of dialkyldichlorostannanes [30]. R_2SnCl_2 has been found to be a very active catalyst for the polymerization of ethylene oxide, the polymerization rate increasing considerably with the length of the alkyl substituent at the tin atom.

A similar dependence is also observed in the polymerization of dialkyl-bis(γ-glycidyloxypropyl)stannanes [30].

On the basis of these results it is suggested that steric factors on the one hand and electron-accepting properties of $R_2Sn<$ groups, varying with the length of the substituents at the tin atom, on the other hand are responsible for the considerable change in the polymerization rate of dialkylepoxystannanes.

In contrast to the above-mentioned organotin epoxide monomers, dialkyl(diphenyl)allyl-2,3-epoxypropylstannanes with the general formula

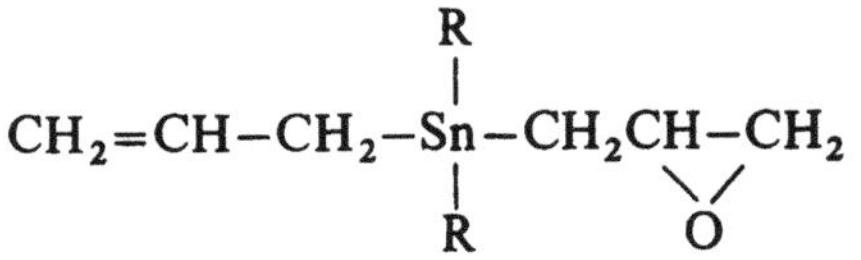

readily undergo spontaneous polymerization. This is probably due to a unique feature in the structure of allylepoxystannanes: the molecule is strongly strained due to π-allyl conjugation with the tin atom, the epoxy group being in a coordination-bound state. The tendency of these monomers to undergo polymerization greatly depends on the length of the alkyl substituents at the tin atom. On transition from CH_3 to C_4H_9, the contribution of the Sn—C bond to conjugation and complexing with the allyl and epoxy groups is reduced. Thus, a high polymerizability of a dimethylstannyl derivative can be expected. In fact, according to the experimental results, it is impossible to isolate dimethyl- and diethylallyl-2,3-epoxypropylstannanes from the reaction mixture since these compounds are immediately converted into a polymeric powder whereas dipropyl- and dibutylstannyl derivatives are actually isolated and identified. The polymerizability of these compounds was found to decrease in the following order:

$$CH_3 > C_2H_5 \gg n\text{-}C_3H_7 > n\text{-}C_4H_9 \, .$$

The analysis of the IR-spectra of the starting monomers and the products resulting from spontaneous polymerization of di-n-propylallyl-2,3-epoxypropylstannane indicates that both the allyl and epoxide group activated by the organotin residue participate simultaneously in the polymerization process. This is confirmed by the disappearance of the absorption bands at 3080 and 1625 cm^{-1} typical of the allyl group and by the appearance of a new narrow band of medium intensity near 3500 cm^{-1} which is characteristic of the stretching vibration of a hydroxy group in the intermolecular-bound state of the macromolecule.

4 Cross-Linking

Organotin oligomers and polymers can be readily cross-linked upon irradiation.

Low-molecular weight organotin compounds are known to easily undergo chemical conversions upon UV-irradiation [1]. However, the photochemistry of organotin polymers is still obscure.

Of great importance for both the formation and photochemical cross-linked of organotin oligomers and polymers is the tin atom in the coordination-bound state.

A considerable viscosity increase in copolymers of tributylstannyl methacrylate with methyl methacrylate, butyl acrylate and styrene upon prolongated storage has been observed and special agents to eliminate this effect have been proposed [108]. It is likely that the destruction of intermolecular coordination complexes formed by involvement of tin and carbonyl groups in comonomer units takes place in this case.

The authors of Refs. [99,109] have found that coatings and films formed from oligoorganoepoxystannanes and organotin copolymers became insoluble on exposure to light. However their organic analogs are quite stable under the same conditions.

Cross-linking of organotin epoxy oligomers was investigated by IR spectroscopy at various stages of photo-oxidation with UV-light [99]. Thus, the change in the shape and shift of maximima of the absorption bands and the appearance of new typical bands in the same or another region, caused by UV irradiation, were controlled in the IR spectrum.

The spectral results showed that exposure of the sample to UV light causes a drastic change in the spectrum of oligoorganoepoxystannanes manifested by a shift of the maximum of the absorption band from 500 (ν_{Sn-C}) to 530 cm^{-1}; the appearance of new bands at 1595 and 1730 cm^{-1} ($\nu_{c=o}$ in R_3SnOOC—), 3450 (ν_{OH}) and 960 cm^{-1}

(ν_{Sn-O}); the disappearance of absorption bands at 1640 ($\nu_{C=C}$), 3080 (δ_{CH} in (δ_{CH} in $\underset{O}{\nabla}$ and $=CH_2$), 1250 and 765 cm^{-1} ($\underset{O}{\nabla}$); lower intensity of bands at 1420 and 1640 cm^{-1} (δ_{CH} in CH_2—Sn) and broadening of the band at 110 cm^{-1} (C—O—C).

It has been suggested that the observed partial shift, i.e. a shoulder of the band at 500 cm^{-1}, and increase of its intensity, the total dissappearance of the band at 500 cm^{-1}, and the appearance of a new band at 530 cm^{-1} due to UV irradiation are caused by a disturbance of the pentacoordination state of the tin atom [99].

It is interesting to note that the most critical spectral changes occur at the initial stages of photo-oxidation. On further exposure of the samples (as a thin film on KBr) to UV light, a certain stabilization of chemical changes in the structure of the oligomers studied takes place along with a decrease in the intensity of the major absorption bands.

Apparently, photo-oxidation processes are dominant during the initial period and lead to the formation of active centers contributing to further cross-linking of macrochains. It can be assumed that methylene groups adjacent to coordination-bound tin atoms are most susceptible to photo-oxidation.

The observed spectral changes suggest the following photochemical cross-linking mechanism for oligoorganoepoxystannanes:

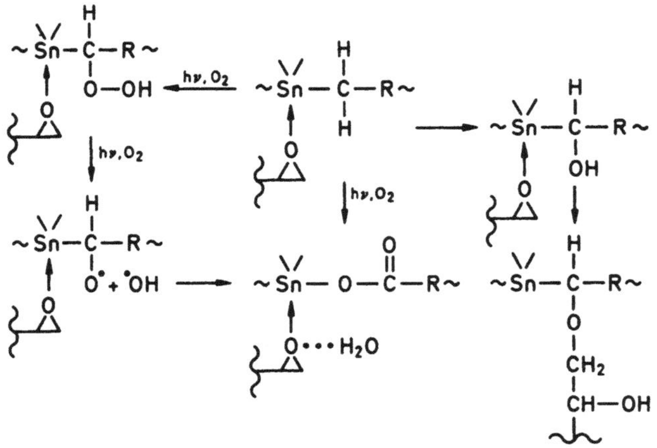

The IR spectra of thin films from an alternating copolymer of tri-n-butylstannyl methacrylate and maleic anhydride at various stages of UV irradiation were also studied [109]. It can be expected that cross-linking results in hindered rotational and oscillatory mobility of the backbone chain and of side chains in the macromolecule. Indeed, a decrease in the peak intensity of most absorption bands is observed in the IR spectra of an irradiated copolymer: 1770 and 1840 cm^{-1} ($\nu_{C=O}$ anhydride), 1406 cm^{-1} (δ_{CH} in CH_2—Sn), 1285 and 1080 cm^{-1} (ν_{C-O-C}), 1115, 850, 750cm^{-1}, etc. At the same time, there is an increase in the intensity of absorption bands near 1720 ($\nu_{C=O}$ in $—COOSnR_3$) and 1580 cm^{-1} (ν_{C-O} in $—C=O ... SnR_3$).

The intensity ratio between typical absorption bands and the least changing band was used to illustrate the changes in the spectra. The change (m) in the absorption bands is mainly observed at the initial irradiation stage (30–40 min). The most marked decrease in the band intensity was found for the carbonyl group in the anhydride unit (1770 cm^{-1}).

The change Δm ($\Delta m = m_0 - m_{30}$) at the initial irradiation stage (for 30 min) can serve as a parameter characterizing photochemical conversions into a copolymer during absorption of radiation energy. From this evidence, the values of Δm for $\nu_{C=O}$, (1770 cm^{-1}) were found as $\Delta m_{30}^{1380} = 0.16$ and $\Delta m_{30}^{1230} = 0.12$. If a sensitizer (3% of 9,10-dibromoanthracene with $E_T = 168$ kJ/mol and $E_s = 294$ kJ/mol) is used [110], the value of Δm_{30} for $\nu_{C=O}$ increase from 0.12 to 0.37 which is probably due to the absorption of visible light by the sensitizer [109].

It is seen from the electron spectrum that a TBSM — MA copolymer absorbs in the UV region within a wide range (190–240 nm) with a maximum at $\lambda = 205$ nm while the sensitizer absorbs in both the UV (250–280 nm, $\lambda = 260$ nm) and 360–420 nm region with two maxima at 380 and 415 nm. Irradiation of the copolymer results in a lower intensity maximum at $\lambda = 205$ nm which indicates the occurrence of photochemical reactions.

A drastic decrease in the relative intensity of the analytical band at 1770 cm^{-1} is also observed when a sensitized polymer film is irradiated by monochromatic light ($\lambda = 405$ nm). The value of Δm_{30} in this case was found to be 0.22 which only slightly differs from that obtained by UV irradiation under similar conditions. It appears that the sensitizer is responsible for photochemical conversions in the visible region [109].

In all cases where thin copolymer coatings were irradiated by light, the exposed portions became insoluble in organic solvents while unexposed coatings were readily washed out by solvents such as acetone, benzene, etc.

The experimental findings of the optical density in the absorption region of the C=O group (1770 cm^{-1} at a layer thickness of $h = 0.87 \times 10^{-4}$ cm), of the molar extinction coefficients of irradiated and non-irradiated copolymer films, and of the intensities of absorbed light (= 405 nm) made it possible to determine the quantum efficiency of C=O group consumption using the known equation

$$\varphi = \frac{\Delta n}{I_n^{abs} \cdot t} = 1.46 \pm 0.2 \text{ mol/Einstein}$$

where $\Delta n = n(C_o - C_t) \times 10^{-3}$ and C_o and C_t = molar concentrations of anhydride units before and after irradiation, respectively, calculated from optical density changes with the formula

$$C^{1770} = D^{1770}/^{1770}\, h\, ;$$

I_n^{abs} = film absorption = $I_p^{abs} - I_k^{abs}$, i.e. the difference of light absorptions on a NaCl glass with and without film, respectively ($I_p^{abs} = I_o - I_p$ and $I_k^{abs} = I_o - I_k$); and t = irradiation time (s).

The light intensity ($\lambda = 405$ nm) was determined actinometrically using the formula

$$I_o = n_{Fe}/\varphi_{405} \cdot t$$

where φ_{405} = quantum efficiency of bivalent iron formation in the actinometer at $\lambda = 405$ nm ($\varphi_{405} = 1.14$ mol/Einstein).

This quantum efficiency of the photochemical conversion of the polymer suggests a complicated mechanism of these reactions in the presence of O_2. On the other hand, the low segmental mobility of macromolecules in contrast to that of the low-molecular weight analogs inhibits peroxide formation and favors other reactions essential for photocross-linking. As might be expected, this results in a higher probability of two quantum processes to occur which is generally observed when polymers are irradiated in an oxygen atmosphere [111].

IR spectra reveal that hydroxy groups are formed near 3500 cm^{-1} (broad band) as cross-linking proceeds. The intensity of this band increases with irradiation time. An increase in intensity of the band at 1580 cm^{-1} ($\nu_{C=O}$ in —C=O ... SnR_3) coinciding with a decrease of the intensity of the 1700 ($\nu_{C=O}$) and 1080 cm^{-1} (ν_{C-O-C}) bands indicates that free carbonyl groups are converted to a coordination-bound fragment. The intensity of the absorption bands at 1640 cm^{-1} relating to the methylene group at the tin atom (δ_{CH_2} in CH_2—Sn) decreases which suggests that the photooxidation reaction of CH_2 groups advances until hydroxy groups have been formed.

It is likely that cross-linking of an organic copolymer proceeds through a stage of excitation and photochemical conversion of photosensitive side-chain organotin fragments containing coordination-bound residues.

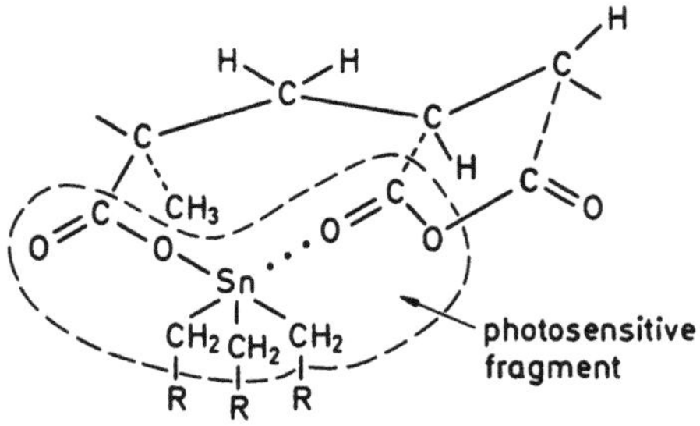

This ensures the necessary ordering of reacting side groups for an elementary act to take place although the copolymer macromolecules exhibit a low mobility [109].

The changes in the IR spectra of the copolymer exposed to UV irradiation suggest the formation of coordination-bound organotin fragments due to complex intermolecular reactions of anhydride and organotin units.

I

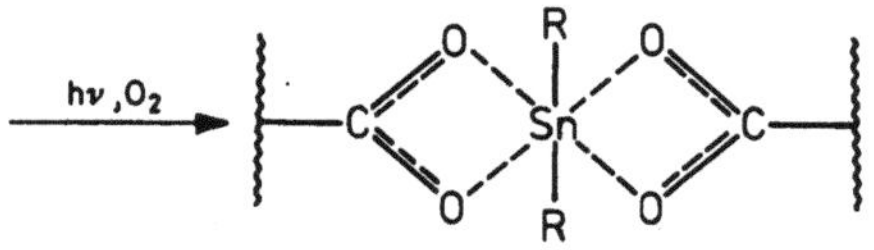

The spectral results can be corroborated by electron microscopy which is capable of tracing the photocross-linking process on a supramolecular level [109].

UV irradiation drastically changes the supramolecular structure. The starting non-irradiated copolymer contains large ordered structures the formation of which is apparently due to the presence of bulky coordination-bound organotin fragments in the macromolecular chain. At the initial irradiation stage the structures increase in volume and turn into mutually bound globular chain-packed forms. This means that the exposure to light results in physical cross-linking, i.e. transition of organotin macromolecules into a more thermodynamically stable state preceding their photocross-linking.

Then, it appears, that photochemical cross-linking of an organotin copolymer is the result of complex supramolecular conversions and intermolecular reactions of anhydride and organotin units involving the formation of transverse coordination-bound organotin carboxylate fragments [109].

It follows from these findings that the simultaneous presence of both anhydride and organotin groups in the copolymer structure and their regular distribution among the macromolecular chain is a prerequisite for photochemical cross-linking of a polymer.

Under similar UV irradiation conditions for the coatings obtained from the solution of model copolymers such as poly(tributylstannyl methacrylate) and maleic anhydride/styrene copolymer, no considerable change was observed in their IR spectra and solubility.

5 References

1. Neumann, W. P.: The Organic Chemistry of Tin, Stuttgart, Ferdinand Enke Verlag 1967
2. Rzaev, Z. M., Bryksina, L. V.: Vysokomol. Soedin. *A16*, 1691 (1974)
3. Rzaev, Z. M., Mamedova, S. G., Rustamov, F. B.: 8th All-Union Chugayev Conf. Complex Compounds, Abstr., p. 335, Moscow 1978
4. Borden, P. G.: Amer. Chem. Soc., Polym. Prepr. *18*, 849 (1977)
5. Rzaev, Z. M.: Chemtech. *1*, 58 (1979)
6. Gielen, M., Nasielski, J.: Rev. Trav. Chem. *88*, 228 (1963)
7. Herber, G. P., Stocker, H. A., Reichle, W. T.: Chem. Phys. *42*, 2337 (1965)
8. Gielen, M., Sprecher, N.: Organometal. Chem. Rev. *1*, 455 (1966)
9. Peddle, G. J. D., Redl, G.: Chem. Commun. 1969, 626
10. Stynes, D. V., Allred, A. L.: J. Am. Chem. Soc. *93*, 2666 (1971)
11. Graddon, D. P.: 3d Internat. Conf. Organometallic and Coordination Chem. Germanium, Tin and Lead, Abstr. p. 80, Dortmund 1980
12. Kumar Das, V. G. et al.: ibid., p. 75, Dortmund 1980
13. Abu-Samn, R. H.: ibid. Abstr., p. 72, Dortmund 1980
14. Tzchach, A., Jurkschart, K.: ibid. Abstr., p. 72, Dortmund 1980
15. Barbieri, R. et al.: Gazz. Chim. Ital. *104*, 885 (1974)
16. Sau, A. C., Carpino, L. A., Holmes, R. R.: J. Organometal. Chem. *197*, 181 (1980)
17. Nasielsky, G. J.: Pure Appl. Chem. *30*, 3 (1972); *16*, 1229 (1972)
18. Sadikh-zade, S. I. et al.: Vysokomol. Soedin. *B13*, 481 (1971)
19. Rzaev, Z. M., Sadikh-zade, S. I.: J. Polym. Sci., Polym. Symp. *42*, 541 (1973)
20. Rzaev, Z. M., Rustamov, F. B., Mamedov, S. M.: First All-Union Conf. Organometallic Chem., Abstr., p. 100, Moscow 1979
21. Rustamov, F. B., Gasanov, K. G., Rzaev, Z. M.: Azerb. Khim. Zh. *2*, 75 (1980)

22. Rzaev, Z. M., Mamedova, S. G., Medyakova, L. V.: Second Intern. Symp. Homogeneous Catalysis, Düsseldorf 1980
23. Rzaev, Z. M., Mamedova, S. G., Rustamov, F. B.: Third Intern. Conf. Organometallic and Coordination Chem. of Germanium, Tin and Lead, Abstr., p. 13, Dortmund 1980
24. Hajiev, T. A., Rzaev, Z. M., Mamedova, S. G.: Vysokomol. Soedin. *A13*, 2386 (1971)
25. Rzaev, Z. M., Mamedova, S. G.: J. Polym. Sci., Polym. Symp. *42* 1397 (1973)
26. Ketelaar, J. A. A.: Rev. Trav. Chem. *71*, 1104 (1952)
27. Gagina, I. A., Trapeznikov, A. A.: Simakov, Y. S.: Zh. Prikl. khim. *53*, 177 (1980)
28. Rzaev, Z. M., Kyazimov, Sh. K., Mamedov, S. M.: Azerb. Khim. Zh. *3*, 85 (1972)
29. Rzaev, Z. M. et al.: Vysokomol. Soedin. *B15*, 853 (1973)
30. Rzaev, Z. M. et al.: Azerb. Khim. Zh. *6*, 51 (1977)
31. Chumayevski, N. A.: Oscillation Spectra of Heteroorganic Compounds with IV B and V B Element, Moscow, Nauka Publishers, 1971, (Russ.)
32. Nakanisi, K.: IR Spectra Structure of Organic Compounds, Moscow, Mir Publishers, 1965, (Russ. Transl.)
33. Bhacca, N., Williams, D.: Application of NMR Spectroscopy in Organic Chem., Moscow, Mir Publishers 1966, (Russ. Transl.)
34. Tor, K. et al.: Tetrahedron Lett. *1964*, 559
35. Rzaev, Z. M. et al.: Third All-Union Conf. on Biologically Active Compositions of Silicon, Germanium and Tin Compounds, Abstr., p. 63, Irkutsk 1980
36. Gusev, A. I. et al.: Third All-Union Conf. on Organic Crystal Chemistry and on the Section of Crystal Chemistry of the Structure and Properties of Polymetallic Organometallic Compounds, p. 9, Gorky 1981
37. Kuzmina, L. G., Struchkov, Y. T.: ibid., p. 10, Gorky 1981
38. Andrianov, K. A.: Polymers with Inorganic Backbones, Moscow, ANSSSR 1962, (Rqss.)
39. Andrianov, K. A., Khananashvili, L. M., Haiduk, J.: Progress in Polymer Chemistry (Rev. by V. V. Korshak), Moscow, Nauka Publishers 1969, (Russ.)
40. Andrianov, K. A.: Methods of Heteroorganic Chemistry. Silicon. Moscow, Nauka 1968, (Russ.)
41. Andrianov, K. A., Khananashvili, L. M.: Technology of Heteroorganic Monomers and Polymers, Moscow, Khimiya Publishers 1973, (Russ.)
42. Rochow, E. G., Hurd, D. T., Lewis, R. N.: The Chemistry of Organometallic Compounds, Moscow, In. Lit. Publishers 1963, (Russ. Transl.)
43. Lukevits, E. L., Voronkov, M. G.: Hydrosilylation, Hydrogermanilation and Hydrostannylation, Riga, Latv. Acad. Aci. 1964, (Russ.)
44. Seifert, D.: Organometallic Compounds with a Vinyl Group, Moscow, Mir Publishers 1964, (Russ. Transl.)
45. Kocheshkov, K. A. et al.: Methods of Heteroorganic Chemistry, Moscow, Nauka Publishers 1968, (Russ.)
46. Kochkin, D. A., Azerbayev, I. N.: Organotin and Organolead Monomers and Polymers, Alma-Ata, Nauka Publishers 1968, (Russ.)
47. Polymer Encylopaedia, Vol. 2, Sovetskaya Enzyklopediya, p. 477, 1974, (Russ.)
48. Korshak, V. V., Polekova, A. M., Suchkova, M. D.: Vysokomol. Soedin. *2*, 12 (1960)
49. Korshak, V. V. et al.: Izv. Akad. Nauk SSSR, Otd. Kihm. Nauk *1959*, 178
50. Minoura, J. et al.: J. Polym. Sci. *4*, 2757 (1966)
51. Plate, N. A. et al.: Vysokomol. Soedin. *A8*, 1890 (1966)
52. Samoilov, S. M. et al.: Vysokomol. Soedin. *B16*, 42 (1974)
53. Zhdanov, A. A., Andrianov, K. A., Fetisov, G. A.: USSR Inventor's Certificate 233336 (1968); Refer. Zh. Khim. 18C236 (1969)
54. Maltseva, Y. N., Zavgorodny, V. S., Petrov, A. A.: Zh. Org. Khim. *39*, 152 (1969)
55. Zavgorodny, V. S., Maltseva, Y. N., Petrov, A. A.: Zh. Org. Khim. *39*, 159 (1969)
56. Makarov, K. A. et al.: USSR Inventor's Certificate 267911 (1970); Refer. Zh. Khim. 30290 (1971)
57. Zavgorodny, V. S. et al.: USSR Inventor's Certificate 284992 (1971); Refer. Zh. Khim. 15C5381 (1971)
58. Makarov, K. A. et al.: Vysokomol. Soedin. *A12*, 1429 (1970)
59. Makarov, K. A., Solovyova, T. K., Nikolayev, A. F.: Izv. Vuz SSSR, Khim. i Khim. Techn. *14*, 763 (1971)

60. Chander, R. H.: Belg.-Ned. Tijdschr. Oppervl. Tech. Metal. *21*, 275 (1977)
61. Henry, M. C., Davidson, W. E.: Organotin Compounds, Vol. 3, p. 975, G. Wiley and Sons, New York 1972
62. Subramanian, R. V.: Gard. Polym.-Plast. Technol. Eng. *11*, 81 (1978)
63. Kochkin, D. A., Terteryan, R. A.: Izv. VUZ SSSR, Khim. Khim. Technol. *17*, 724 (1974)
64. Kushlefsky, B. G.: US Patent 4058544 (1977); Refer. Zh. Khim. 9C, 334 (1978)
65. Mufti, A. S., Poller, R. C.: J. Chem. Soc. *1967*, 1362
66. Kochkin, D. A. et al.: Coll. Synthesis of Compounds and Physicochemical Analysis, p. 12, Kalinin, KGU 1973
67. Pawlowa, O. W., Siwergin, K. M.: Plaste Kautschuk *24*, 669 (1977)
68. Maltenieks, O. J.: US Patent 3721690 (1973); Refer. Zh. Khim. 5C, 469II (1974)
69. Matsushita, T., Suzuki, K., Hasegouva, S.: US Patent 3705054 (1972); Refer. Zh. Khim. 3C, 454 (1974)
70. Wende, A., Becker, R.: FRG Patent 1158974 (1963); C.A.60 9311 (1964)
71. Tomita, J.: US Patent 3340212 (1967); Refer. Zh. Khim. 18C 653 (1969)
72. Kauto, H. et al.: Jap. Patent 33489 (1972); Refer. Zh. Khim. 10C 57811 (1973)
73. Iwamoto, N., Ninagawa, A., Mastude, H.: J. Chem. Soc. Japan, Ind. Chem. Soc. *72*, 1847 (1969)
74. Subramanian, R. V., Garg, B. K., Corredor, J.: Am. Chem. Soc., Polym. Prepr. *18*, 850 (1977)
75. Garg, B. K., Corredor, J., Subramanian, R. V.: J. Macromol. Sci. *A11*, 1567 (1977)
76. Zabotin, K. P., Malysheva, L. V.: Proceedings Chemistry and Chemical Technology. Physico-chemical Foundations of Synthesis and Processing of Polymers Vol. 1, p. 474, Gorky, GGU 1973, (Russ.)
77. Zabotin, K. P. et al.: Transactions: Chemistry and Chemical Technology, Vol. 2, p. 24, GGU, Gorky 1974
78. Zabotin, K. P. et al.: Transactions: Chemistry and Chemical Technology, Vol. 2, p. 27, GGU Gorky 1974
79. Kanai, J. et al.: Jap. Patent 50 33894 (1975); Refer. Zh. Khim.2T67711 (1977)
80. Yamada, B., Yoneno, H., Otsu, T.: J. Polym. Sci. *8*, 2021 (1970)
81. Considene, W. J., Reifenberg, G. H.: US Patent 3345069 (1969); Refer. Zh. Khim. 17C 383 (1970)
82. Tsvetkov, O. N., Samoilov, S. M., Monastyrsky, V. M.: Proc. Research Inst. for Petroleum Processing 1976, 15, 100 (Russ.)
83. Lebedev, N. K., Lebedev, B. V., Bryuchanov, A. N.: Uch. Zap. Gorkovskogo Universiteta, Vol. 47, p. 1, 1976
84. Organometallic Polymers, Symp. New Orleans, March 20–22, 1977, Carraher, Ch. E. (ed.), New York, Academic Press 1978
85. Stanczyk, W., Kawalski, I., Chojnowski, I.: Polym.-Tworzywa Wielkoczasteczkowe *24*, 111 (1979)
86. Kawalski, I., Stanczyk, W., Chojnowski, I.: Polym.-Tworzywa Wielkoczastecrowe *24*, 185 (1979)
87. Khalilova, Z. I., Agayev, U. Kh., Rzaev, Z. M.: Azerb. Khim. Zh. *2*, 68 (1974)
88. Rzaev, Z. M. et al.: VIIth IUPAC Microsymp. on Macromolecules. Poly(vinyl chloride), its Formation and Properties, Abstr., p. B1, Prague 1970
89. Ham, D.: Copolymerization, Moscow, Khimiya Publishers 1971, (Russ. Transl.)
90. Hidefumi, H.: J. Macromol. Sci. *A9*, 883 (1975)
91. Georgiyev, G. S., Golubev, V. B., Zubov, V. P.: Vysokomol. Soedin. *A20*, 1608 (1978)
92. Rzaev, Z. M. et al.: Azerb. Khim. Zh. *2*, 80 (1973)
93. Dev, P. C., Samui, A. B.: Angew. Makromol. Chem. *80*, 137 (1979)
94. Smirnova, L. A. et al.: Vysokomol. *A22*, 2137 (1980)
95. Rzaev, Z. M., Bryksina, L. V., Jafarov, R. V.: Vysokomol. Soedin. *A17*, 2371 (1975)
96. Rzaev, Z. M., Halilova, Z. I.: XXIIIth IUPAC Intern. Symp. on Macromolecules, Prepr. of Papers, p. 123, Madrid 1974
97. Korshak, V. V.: Progress of Polymer Chemistry, Moscow, Nauka Publishers 1965 (Russ.); Vysokomol. Soedin. *A18*, 1443 (1976)
98. Rzaev, Z. M., Rustamov, F. B.: Vysokomol. Soedin. *B19*, 576 (1977)
99. Rzaev, Z. M. et al.: Azerb. Khim. Zh. *4*, 85 (1980)

100. Rzaev, Z. M. et al.: Transactions of Kalinin State Univ. in: Structure of Compounds and Physico-chemical Analysis, p. 95, Kalinin 1973
101. Rzaev, Z. M. et al.: Plastmassy *3*, 66 (1975)
102. Rzaev, Z. M. et al.: Izv. VUZ SSSR, Chem., Chem. Technol. *18*, 77 (1975)
103. Rustamov, F. B. et al.: Industrial Branch Coordination Conf. on Epoxy Resins and Materials on their Basis, Abstr., p. 5, Donetsk 1975
104. Mamedova, S. G. et al.: Studies on the Synthesis of Polymer and Monomer Products, p. 188, Baku, Elm Publishers 1977
105. Mathuda, S. M. et al.: J. Chem. Soc. Japan, Ind. Chem. Sec. *71*, 2054 (1968)
106. Marcovitz, M., Kohn, L. S.: US Patent 3728306 (1973); Refer. Zh. Khim. 4C 581 (1974)
107. Go, T., Caiti, T.: US Patent 3745132 (1973); Refer. Zh. Khim. 9C 468 (1974)
108. Atherton, D., Verborgt, I., Winkeler, M. A. M.: J. Coat. Technol. *51*, 88 (1979)
109. Rzaev, Z. M. et al.: Vysokomol. Soedin. *B22*, 831 (1980)
110. Gordon, A., Ford, R.: The Chemist's Companion, p. 373, Moscow, Mir Publishers 1976, (Russ. Transl.)
111. Encyclopaedia of Polymers, Vol. 3, p. 771 Sovetskaya Entsiklopediya, 1977

Bei Fragen zur Produktsicherheit wenden Sie sich bitte an:
If you have any questions regarding product safety,
please contact:

Walter de Gruyter GmbH
Genthiner Straße 13
10785 Berlin
productsafety@degruyterbrill.com